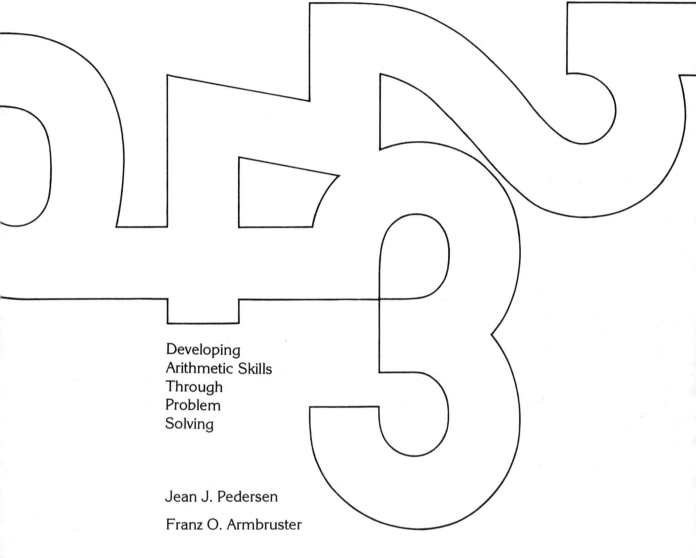

A NEW TWIST

Developing
Arithmetic Skills
Through
Problem
Solving

Jean J. Pedersen

Franz O. Armbruster

Addison-Wesley Publishing Company
Menlo Park, California • Reading, Massachusetts
London • Amsterdam • Don Mills, Ontario • Sydney

Cover Design, Artwork, and Book Design by Edidt Geever

This book is in the Addison-Wesley Innovative Series.

Copyright © 1979 by Addison-Wesley Publishing Company, Inc.
Philippines Copyright 1979.
All rights reserved.
Printed in the United States of America. Published
simultaneously in Canada.

ISBN-0-201-05712-3
DEFGHIJK-EB-8987654

To
George and Stella Pólya

CONTENTS

Introduction	vi
Skills Chart	ix
Are You an Einstein?	1
Special Numbers	3
Almost Perfect	5
Palindrome	7
Four Fours	9
Mean Numbers	11
Logothetti's Way	13
Oh, No!	15
Pan-Numerics	17
Card Squares	19
Ker-Plunk	21
Indianapolis 500	23
Permuprime Pairs	25
Alpha-Symmetric	27
Twentieth Century Palindromes	29
The Die Is Cast	31
Family Harmony	35
Goldbach's Conjecture	37
×Lasrever×	41
Factor Polygons	43
+Lasrever+	45
Kaprekar's Constant	47
Digitadition Sequences	49
Crittenden's Letter	53
Special Areas	55
Special Volumes	57
Rapid Calculations	59
Power	61
𝄞 Albrecht Dürer	63
Digital Roots	67
Grandfather's Models	71
No Duplicates	75
Mental Lightning	77
Columbus Day Discovery	79
Community Property	83
Dots	85
Clocks	87
Domino Arithmetic	93
Odd Sums	97
Glossary	100

INTRODUCTION

WHAT SHOULD WE TEACH?

We think George Pólya answers that question better than anyone we know:

> A great discovery solves a great problem but there is a grain of discovery in the solution of any problem. Your problem may be modest; but if it challenges your curiosity and brings into play your inventive faculties, and if you solve it by your own means, you may experience the tension and enjoy the triumph of discovery. Such experiences at a susceptible age may create a taste for mental work and leave their imprint on mind and character for a lifetime.
>
> Thus, a teacher of mathematics has a great opportunity. If he fills his allotted time with drilling his students in routine operations he kills their interest, hampers their intellectual development, and misuses his opportunity. But if he challenges the curiosity of his students by setting them problems proportionate to their knowledge, and helps them to solve their problems with stimulating questions, he may give them a taste for, and some means of, independent thinking.*
>
> George Pólya

Find the quotients and remainders. Check your work.

- A. 8 ÷ 236
- B. 19 ÷ 158
- C. 38 ÷ 1420
- D. 41 ÷ 5620
- E. 35 ÷ 3163
- F. 412 ÷ 867
- G. 806 ÷ 4176
- H. 814 ÷ 6172
- I. 74 ÷ 4706
- J. 82 ÷ 9162
- K. 132 ÷ 4799
- L. 78 ÷ 4713
- M. 46 ÷ 3216
- N. 52 ÷ 3477
- O. 402 ÷ 30761
- P. 861 ÷ 72468

Task A

$2 \times 12 = 24$
12 backward -21
difference 3 is a prime
therefore 12 is **interesting**

$2 \times 112 = 224$
112 backward 211
difference 13 is a prime
therefore 112 is **interesting**

$2 \times 36 = 72$
36 backward 63
difference 9 is a square
therefore 36 is **interesting**

Find more numbers that are interesting in one way or another.

Task B

---

*George Pólya, **How to Solve It,** Princeton University Press, Princeton, N.J., 1971.

Almost everyone agrees that some degree of proficiency in arithmetic is desirable. Many reasonable young people confronted with pages of drill problems might think of them as they do vegetables—they know they are good for them, but they don't **like** them anyway.

Look at tasks A and B and answer the following question.

Which of the two held your attention the longest? We don't think children are basically different from adults in their approach to rote activities. We believe people of any age would rather answer a challenging question than repeat an algorithm with the only goal being to finish the page.

This book was written to provide interesting questions that teachers could ask students about numbers. We've tried to make the questions amusing enough to inspire students to find the answers—become involved in doing computations—and eventually become more proficient at arithmetic.

We like hand calculators and have found them to be a valuable motivational tool for the students we deal with. Accordingly, we let our students use them. But, situations do exist in which the use of a calculator is not beneficial for beginning students. You will know what's best for your students. The questions in this book will be useful to students both with and without a calculator. The principal difference will be the results, since students with hand calculators generally experiment with larger numbers. The important thing is that they experiment with numbers, large or small.

We feel that when students finish school they ought to be able to:

1. handle the number facts to a degree that will enable them to do simple computations involving addition, subtraction, multiplication, and division (preferably with whole, fractional, and decimal numbers);
2. read, assimilate, and interpret the daily newspaper well enough to function intelligently as members of their community;
3. socialize to the extent that they do not burden society;
4. exercise problem-solving skills to diagnose their own strengths and weaknesses and thereby gain the best mix of employment and recreation, attaining personal satisfaction and fulfillment.

We have long felt that the study of mathematics can, and should, contribute to the objectives in 1 and 4.

Unfortunately, the study of mathematics has been and might continue to be a source of confusion for some and frustration for many. We hope our book will reduce the confusion and alleviate some of the frustration.

The book provides student pages posing problems designed to motivate students to do lots and lots of arithmetic drill. The symbol in the pages' upper right-hand corner shows at a glance what materials are required and skills practiced.

The student pages may be reproduced for classroom use.

Included with each student page is a teacher's page, containing additional hints, suggestions, variations, and, in some cases, full solutions.

We have tried to concentrate on the four fundamental operations, we have not attempted to cover the whole elementary arithmetic curriculum. We haven't focused on sixth graders or tenth graders, or fifteen-year-olds, or boys or girls. What we've tried to do is present problems that will challenge and stimulate **people** to think in a way that will produce computational practice.

The skills chart may help you select activities that meet the needs of your students.

SKILLS CHART

Page	TITLE	HC	Whole numbers +	−	×	÷	Fractions	Decimals	Other
	ARE YOU AN EINSTEIN?	*	X	X					
	SPECIAL NUMBERS	*	X	X	X	X			
	ALMOST PERFECT	*	X	X	X	X			
	PALINDROME	*	X						
	FOUR FOURS	*	X	X	X	X	X	X	powers, roots, etc.
	MEAN NUMBERS	*	X			X	X		averages
	LOGOTHETTI'S WAY						X		algebra (optional)
	OH, NO!	*	X	X	X	X	X	X	algebra (optional)
	PAN-NUMERICS	*	X	X	X	X	X	X	
	CARD SQUARES		X						squares, logic
	KER-PLUNK				X		X		induction, deduction
	INDIANAPOLIS 500	*	X	X					
	PERMUPRIME PAIRS				X	X			primes, squares, ...
	ALPHA-SYMMETRIC		X	X	X	X			symmetry
	TWENTIETH CENTURY PALINDROMES				X	X			special numbers
	THE DIE IS CAST		X		X				induction, deduction
	FAMILY HARMONY	*	X		X	X			
	GOLDBACH'S CONJECTURE	*	X						primes
	× LASREVER ×	*	X		X	X			algebra (optional)
	FACTOR POLYGONS	*			X	X			factorization
	+ LASREVER +	*	X			X			algebra (optional)
	KAPREKAR'S CONSTANT	*	X	X	X				
	DIGITADITION SEQUENCES		X						
	CRITTENDEN'S LETTER		X	X	X	X			palindromes
	SPECIAL AREAS		X		X				geometry
	SPECIAL VOLUMES		X		X				geometry
	RAPID CALCULATIONS		X		X				algebra (optional)
	POWER		X		X				exponents
	ALBRECHT DÜRER		X						symmetry, art, logic
	DIGITAL ROOTS		X			X		X	
	GRANDFATHER'S MODELS		X	X	X	X		X	checking computations
	NO DUPLICATES	*	X	X	X				
	MENTAL LIGHTNING		X		X				"showing off"
	COLUMBUS DAY DISCOVERY		X	X	X	X			logic, deduction
	COMMUNITY PROPERTY		X	X	X	X			induction
	DOTS				X		X	X	geometry, area
	CLOCKS	*	X	X	X	X	X	X	angular measure
	DOMINO ARITHMETIC		X		X	X	X	X	logic, deduction
	ODD SUMS		X	X					

*Topics for which the hand calculator may be useful.

ix

A LOT (of student response)

FOR A LITTLE (planning on your part)

You won't find any remarks in the exercises about the time required for completion. This book is intended as a supplement to your regular textbook; one of the topics can be given to your students once a week. Let them take as much time as they like to work on the problems when they are free. During those free times they might wish to work on this week's topic, or one from a previous week. They will be finding answers to some of the problems for weeks after the exercise is first introduced. You can encourage this long-range activity in several ways.

One approach is to put the results on a bulletin board and continually post the latest answer to each problem. Leave the answers posted, even if you have to put something else on top to save space. Some students may want to leaf through previous results when they get an unusual idea for solving an old problem. You may have other ideas for keeping the problems and results constantly visible. Try them—your students will probably catch your enthusiasm.

Whatever method you choose, let your genuine surprise and pleasure show when you praise or congratulate your students for having found answers, clues, hints, or solutions. When you do this frequently and sincerely, your students will not wonder "Who cares?"—they will know that **you** care. Your interest in the student's success is a powerful stimulus; for some students it may be enough to change their lives. You can hardly hope to do more!

ANSWERS OR SOLUTIONS?

What's the difference between an answer and a solution? Suppose the problem is: Find a prime number between 100 and 1000. An **answer** would be any of the prime numbers bigger than a hundred and smaller than a thousand; a **solution** would be a procedure that finds all those numbers without a lot of trial and error. Usually a solution is not sought or found until a lot of answers are already available. At this point, someone spots a pattern among the answers, tries a hypothesis, modifies it to fit all cases, and then proves it for all numbers.

We've included answers to many of the exercises—in some cases, many answers. Solutions exist for a few exercises, but for many of them there are no known solutions, only answers. In most cases enough answers are given so that you can check the results of your students' work. Of course, you need not know the solution or even all the answers; you only need to understand the problem well enough to recognize an acceptable answer. The teacher pages are designed to help you in this regard.

Probably your students will find more answers than we've included—possibly they will find new solutions. We hope so!

"WHAT GOOD IS IT?" and "WHO CARES?"

For many teachers one of the hardest problems about teaching mathematics is answering the question "What good is it?" Of course, that question rarely comes up if the kind of mathematics taught is needed almost every day. But everyday life is structured so that a person can actually live quite comfortably without ever doing any mathematics or arithmetic. Surprising? Yes! But this is true, especially in the more affluent areas. Ask how much, then write a check; hand a person your credit card, and sign your name. If the check bounces, you put more money in the bank. When the bill arrives from the credit card company, you pay it. You buy something and the salesperson reads your change on the cash register; you fill your car with gas and the pump automatically computes the price. You need not actually do any arithmetic to survive, provided your income is higher than your expenses.

A very real difficulty exists, then, in trying to honestly answer the question "What good is it?" If the students who ask the question are serious they probably won't want to try to understand your answer, perhaps the best you can hope to do is, as Pólya suggests, keep them so "amused and challenged" that they will have no opportunity to ask the question.

We have tried hard, drawing on our experience, knowledge, and intuition, to offer something realistic in the way of challenging problems for students (notice the absence of the word "relevant").

We hope that some students who otherwise might ask "Who cares?" will, instead, get involved in the challenge of problem-solving. They may end up saying, "I don't know what good it is, but it's interesting, so I'll try it!"

We've tried to make every exercise lead to some important branch of mathematics. In some cases, we show a connection between mathematics and other academic disciplines. In every case, we've tried to give maximum opportunity for problem-solving.

Some exercises are particularly rich and will lead the inquisitive student to areas of mathematics filled with unexpected applications. Others are somewhat shallow in that respect and are an end in themselves. We regret this, but have been continually surprised that those shallow exercises catch some of our students' imaginations, and provide them with hours of arithmetic practice they would not otherwise have had. We suggest that you consider your objective, then select those exercises that you think will motivate your students to do the kinds of practice they need.

Jean J. Pedersen
Franz O. Armbruster

ARE YOU AN EINSTEIN?

Albert Einstein was bright, even as a child. It is reported that when he was six years old someone asked him if he could place plus and minus signs between the digits

1   2   3   4   5   6   7   8   9   =

so that the answer on the right would be 100. The digits had to appear in consecutive order.

Allegedly, Einstein wrote out

1 + 2 + 3 − 4 + 5 + 6 + 78 + 9 = 100,

which was a correct solution.

Can you find any **other** solutions?

ARE YOU AN EINSTEIN?

How do we know this story? Do you suppose it is because Einstein remembered a special adult, or because some special adult remembered Einstein?

Suppose that you, right now, have a young Einstein near you. Will you be able to remember him or her twenty or thirty years from now? Some people don't think of children as individuals—only as problems.

If you do have an Einstein near you—**pay attention!** Someday you can say "I remember when so-and-so used to do such-and-such, and that was the only youngster I ever knew who did that." Try to foster their special talents.

It is interesting to ask children if they know who Einstein was and discover how they react to this story about him. Try it, then find out if any of the young people you know have ever done anything that surprised adults.

SOME SOLUTIONS
(Others may be possible)

$123+45-67+8-9=100$
$12+3-4+5+67+8+9=100$
$123-45-67+89=100$
$12-3-4+5-6+7+89=100$
$-1+2-3+4+5+6+78+9=100$
$123-4-5-6-7+8-9=100$
$1+2+34-5+67-8+9=100$
$1+23-4+5+6+78-9=100$
$1+23-4+56+7+8+9=100$
$12+3+4+5-6-7+89=100$
$1+2+3-4+5+6+78+9=100$

...

AH HAH! Some solutions use the minus sign as a "take away," and some use it to indicate a negative number. That's probably part of the reason that negative numbers are difficult for some people.

Do any other mathematical symbols have this dual character?

There certainly are lots of **words** that involve this kind of difficulty. Right? Or should we say "Write"?

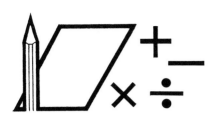

SPECIAL NUMBERS*

A_1 B_2 C_3 D_4 E_5 F_6 G_7 H_8 I_9

Two is a special even number because

$$T_{20} + W_{23} + O_{15} = 58.$$

Both two and 58 are even.

Eleven is a special odd number because

$$E_5 + L_{12} + E_5 + V_{22} + E_5 + N_{14} = 63.$$

Both eleven and 63 are odd.

How many special odd numbers can you find?

How many special even numbers can you find?

There exists a seven-letter English word with a value of 21. Can you find it?

J_{10} K_{11} L_{12} M_{13} N_{14} O_{15} P_{16} Q_{17}

Z_{26} Y_{25} X_{24} W_{23} V_{22} U_{21} T_{20} S_{19} R_{18}

*Special because of the sum of the letter values.

SPECIAL NUMBERS

Try to find other numbers that are special for some reason. Or, look at words. For example, can you guess why the word "prime" is special in some way?*

Almost any category you can think of will have words which, when converted to their numeric values, will be interesting. Try it. Find the sum of the numeric values of the letters in your own name, or birth month. You may find an interesting number related to yourself. This is the idea on which the pseudoscience of numerology is based. If numerologists can find relationships, why can't you? (Anyway, you'll do some addition, so it won't be a total loss!)

Tell your kids this—let them in on the secret of thinking creatively. Remember, the situation that develops creativity is the one that rewards curiosity rather than conformity.

ZERO = 64	FIVE = 42	TEN = 39	FIFTEEN = 65
ONE = 34	SIX = 52	ELEVEN = 63	SIXTEEN = 96
TWO = 58	SEVEN = 65	TWELVE = 87	SEVENTEEN = 109
THREE = 56	EIGHT = 49	THIRTEEN = 99	EIGHTEEN = 73
FOUR = 60	NINE = 42	FOURTEEN = 104	NINETEEN = 81

The word **cabbage** has a value of 21. Does this happen for any other seven-letter English word?

---

*P + R + I + M + E = Prime
16 + 18 + 9 + 13 + 5 = 61

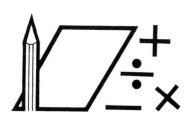

ALMOST PERFECT

Number	Proper Divisors*
2	1
3	1
4	1, 2
5	1
6	1, 2, 3
7	1
8	1, 2, 4
9	1, 3
10	1, 2, 5
11	1
12	1, 2, 3, 4, 6
13	1
14	1, 2, 7
15	1, 3, 5
16	1, 2, 4, 8
17	1
18	1, 2, 3, 6, 9
19	1
20	1, 2, 4, 5, 10

$6 = 1 + 2 + 3$
6 is a **perfect** number.

$12 \neq 1 + 2 + 3 + 4 + 6 = 16$
but
$12 = 1 + 2 + 3 + 6$

12 is **almost** perfect.

How many almost perfect numbers can you find?

Perfect numbers are harder to find. Can you find any of those?

*All the divisors of a number, except the number itself

ALMOST PERFECT

Look at six on the chart. Observe that the sum of **all** the proper divisors of six add up to six. The Greeks were so impressed with this property that they named all numbers possessing that property **perfect** numbers.*

We already know that perfect numbers are hard to find. As you can see, twelve is not perfect because the sum of its proper divisors is sixteen; but, if you take out the four, the sum of the remaining proper divisors does equal twelve. So, we call twelve an **almost perfect** number. More formally, in an almost perfect number the sum of all **but one** of its proper divisors equals that number. To discover if a number is almost perfect, you first find all its proper divisors, then add them and look for **just one** number you can delete to make the total equal the number you started with. This additional flexibility makes almost perfect numbers easier to find.

Challenge your students to be explorers—taking an expedition through the forest of counting numbers to see how many almost perfect counting numbers they can find.

Are there any **odd** almost perfect numbers? No one has ever found any odd perfect numbers!

---

*Early mathematicians spent a good deal of effort trying to find **all** of the perfect numbers. For example, Euclid and Leonard Euler, between them, managed to show that an **even** number is perfect if and only if it is of the form $2^n(2^{n+1}-1)$, where $(2^{n+1}-1)$ is a prime. Thus when $n=1$, $2^1(2^2-1)$ is perfect because $2^2-1=3$ is prime; when $n=2$, $2^2(2^3-1)$ is perfect, because $2^3-1=7$ is prime; when $n=3$, $2^3(2^4-1)$ is **not** perfect, because $2^4-1=15$ is **not** prime.

PALINDROME
(*PAL*-IN-DROME)

How many steps?

Does the Staircase ever end?

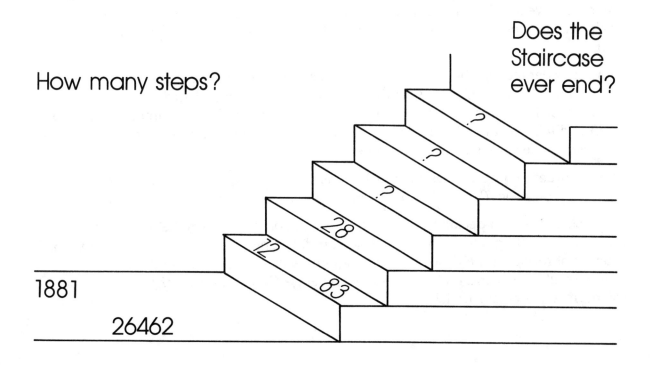

1881

26462

```
  12 ⎞  One step to
+ 21 ⎦  palindrome
  33
```

```
  38 ⎞  One step to
+ 83 ⎦  palindrome
 121
```

```
  28 ⎞  Two steps to
+ 82 ⎦  palindrome
 110 ⎞
+011 ⎦
 121
```

Can you find numbers to put on the third, fourth, and fifth steps?

What is the highest step you can find a number for?

PALINDROME

A number is a palindrome if it reads the same backward and forward. All single digit numbers are palindromes, so we exclude them from the problem. The smallest multidigit number that is palindromic is eleven, the next is twenty-two, then thirty-three, and so on. We say that such numbers have zero steps to palindrome, because you don't have to modify them to make them palindromic. They are palindromic at the start.

Now, consider the number 12. Is it palindromic? No. Reverse the digits (21) and add that to the original number (33). Is the result of the addition palindromic? Yes. Conclusion: The number 12 takes one step to palindrome. Try the number 38. Add 38 to 83 and the result is 121. Is this last number a palindrome? Yes. Therefore, thirty-eight takes one step to palindrome.

Look at twenty-eight—it takes twice using this process to get a palindrome (see the illustration on page 7). Conclusion: Twenty-eight takes two steps to palindrome.

The task, for all who decide to accept it, is to find some numbers that take three steps, four steps, etc., to get to palindrome. Can you draw any general conclusions about this process of reversing the digits, then adding the result to the original number? Will it always result in a palindromic number (eventually)? Can you prove it? How would you try to prove it, or disprove it?

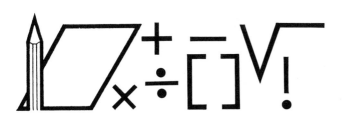 

FOUR FOURS

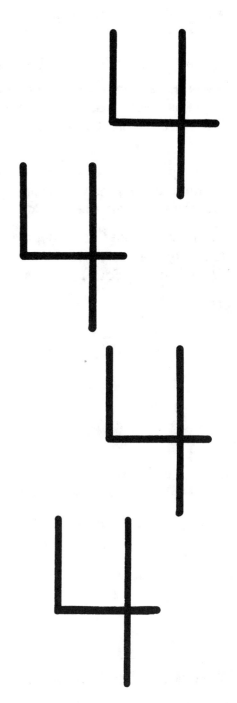

$$\frac{4+4}{4+4} = 1$$

$$\frac{4 \times 4}{4+4} = 2$$

$$\frac{4+4+4}{4} = 3$$

? = 4

Use four 4's

? = 5

? = 6

? = 7

? = 8

⋮

How far can you go?

FOUR FOURS

Notice that if you add 4 + 4 + 4 + 4, you'd get sixteen. But we don't want you to get sixteen yet. We want you to get five, then six, and finally as many numbers as you can in sequence. Use any operations you need, but only four fours can be visible on the left side of the equation.

The most common misconception will be that there is only one right answer for each of the numbers. This, of course, isn't so. It is interesting to see how many different ways you can use the four fours to obtain expressions equivalent to each of the numbers.

It may be helpful to allow other operators (not too early—first teach +, −, ×, ÷).*

If you use **square roots,** you can write $\sqrt{4} = 2$.
If you use **factorials,** you can use $4! = 1 \times 2 \times 3 \times 4 = 24$.
If you use the **greatest integer** function you can write
$$[4.4] = 4 \text{ or } [-4.4] = -5.$$

Make up the rules to suit the participants. Don't underestimate the ability of young children to grasp new ideas like factorials, square roots, etc., especially when they have problems in which they would be helpful.

For more activities of this type, see **Recreational Mathematics Magazine** (January–February 1964), Marjorie Bicknell and Verner E. Hoggatt, Jr., "64 Ways to Write 64 Using Four Fours."

---

*See glossary for definitions and examples of various operators.

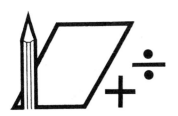

MEAN NUMBERS

138 is a 3-digit number
and $\frac{1+3+8}{3} = 4$ (a whole number).

127512 is a 6-digit number
and $\frac{1+2+7+5+1+2}{6} = 3$ (a whole number).

127 is a 3-digit number
and $\frac{1+2+7}{3} = \frac{10}{3}$ (not a whole number).

Call 138 and 127512 **mean** numbers.

Six of these numbers are mean numbers. Can you find them?

```
    43              32          421           702
        2824              4543
                                          4171
                42                   17
    1113             327
```

How many 2-digit mean numbers can you find? Is there a pattern to their mean values?

(138 has a mean value of 4.)

MEAN NUMBERS

In this context mean, of course, doesn't mean **mean,** mean means **average.** And average doesn't mean mediocre, average means the sum divided by the number of terms.

For the sake of this exercise, **mean numbers** are defined as those numbers in which the sum of the digits in the number divided by the number of digits is a whole number. For example, take a four-digit number. Add the sum of the digits. Divide that result by four. Did the answer come out even (without a remainder)? If it did, you started with a mean number.

Again, take a three-digit number. Add the digits and divide by three. Did the division have a zero remainder? If it did, you had a three-digit mean number.

Look at the two-digit number 28. We say 28 is a mean number because 2 plus 8 is 10, and 10 divided by 2 is 5, with no remainder. Now, look at 127; $1+2+7=10$ and 10 divided by 3 is $3\frac{1}{3}$ which is not a whole number. So, 127 is **not** a mean number.

Explain this to your students, then have them check the numbers shown in the rectangle to see which are mean numbers. If they can correctly identify the six mean numbers they are ready to proceed to the questions about mean numbers. You might ask them to guess how many three-digit mean numbers there are (and verify their guesses, of course).

Now, if you have any students who feel they can "solve-any-problem," have them figure out what percentage of the numbers between zero and one million are mean numbers! (That's a **mean** question.)

LOGOTHETTI'S WAY

Some people add fractions this way:

$$\frac{5}{7}+\frac{2}{3}=\frac{3}{3}\times\frac{5}{7}+\frac{2}{3}\times\frac{7}{7}=\frac{3\times 5}{3\times 7}+\frac{2\times 7}{3\times 7}=\frac{3\times 5+2\times 7}{3\times 7}=$$

$$\frac{15+14}{21}=\frac{29}{21}$$

Professor David Logothetti thinks of adding fractions like this:

$$\begin{array}{c|c|c|c} & 5 & 7 & \\ \hline 2 & + & 2\times 7 & 3\times 5+2\times 7 \\ \hline 3 & 3\times 5 & 3\times 7 & 3\times 7 \end{array}$$

Only he writes it like this:

$$\begin{array}{c|c|c|c} & 5 & 7 & \\ \hline 2 & + & 14 & 29 \\ \hline 3 & 15 & 21 & 21 \end{array}$$

Is his method **really** different from the first method?

Do you believe it will **always** work?

....

Try several cases both ways and decide which method you prefer.

LOGOTHETTI'S WAY

Adding fractions in the usual way is a worthwhile exercise because it explains **why** it is done the way it is. However, Professor David Logothetti, of the University of Santa Clara, has devised a shorter way of writing down the computation (see page 13).

We recognize that this process has an inherent danger, since some students may end up knowing how to add fractions without really knowing why they do it that way. It should **not** be used the first time the addition of fractions is discussed.

However, when you are covering this material for the second or third time we think this method has certain advantages. For the students who already know how to add fractions, it is something new to think about; you can give them some of the advanced problems below. For those students who did not understand adding fractions the first time, perhaps this will encourage them to try again. Students who failed with some particular mathematical algorithm have a strong tendency to dismiss it as "something I'll never be able to do" whenever it is brought up again. Consequently, repeating exactly what was done last year is likely to start those students doodling or daydreaming.

Several elementary and junior high school teachers have indicated that when they used this approach many of the previous "underachievers" became quite adept at adding fractions. Even though some didn't understand why it worked, they were delighted to **finally** be able to perform those computations.

The method can be extended to accommodate several terms as shown below.

$$\tfrac{1}{3} + \tfrac{1}{2} + \tfrac{2}{5}$$

	1	/2		2	/5	
1	+	2	5	+	25	37
3	3	6	6	12	30	30

ADVANCED PROBLEMS

How many ways can Professor Logothetti fill in this array and get an answer less than or equal to one?

	/	
+		
	144	144

Vary the Number here.

Show that Professor Logothetti's method will always work by completing the following computations and comparing the answers.

$$\frac{a}{A} + \frac{b}{B} = \frac{B}{B} \times \frac{a}{A} + \frac{b}{B} \times \frac{A}{A} =$$

	a	/ A
b	+	b × A
B	B × a	B × A

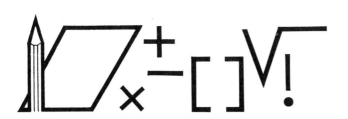 

On April Fool's Day Mr. Carman's students decided to play a trick on him. They handed in calculations that looked like this:

$\dfrac{1\cancel{6}}{\cancel{6}4} = \dfrac{1}{4}, \quad \dfrac{4\cancel{9}5}{9\cancel{9}0} = \dfrac{45}{90},$

······· Cancel any number on top and bottom.

$3^4 \cdot 425 = 34425,$
$8^2 - 2^2 = 82 - 22,$
$6^2 - 4^2 = 62 - 42,$

······· Exponents don't mean anything—the number just slipped up.

$(7 + \dfrac{3}{7})(4 - \dfrac{3}{13}) = 7 \cdot 4,$
$(6 + \dfrac{1}{4})(5 - \dfrac{1}{5}) = 6 \cdot 5,$

······· Fractions are troublesome—so just ignore them.

The students called their computations **misteaks,** incorrect operations that give correct answers.

How many more misteaks can you make up?

OH, NO!

We have found this activity is best suited to students who already have a good understanding of the correct ways to simplify fractions and handle exponents. It may give them something interesting to think about while you help other students with more fundamental concepts. We would **not** use it for students who have not yet mastered these ideas. You must employ a good deal of tact to convince the more advanced students that they should **not** discuss these incorrect ideas with the other students. If, however, you can convince them that it might be useful for them to **help** their classmates understand the correct ideas you may be able to interest them in finding out what kinds of **mistakes** (honest errors) students make—and those might give them ideas for their own **misteak** problems. At any rate, proceed with caution!

The article, "Mathematical Misteaks," by Professor Robert A. Carman of Santa Barbara City College* is loaded with good examples of **misteaks** and how to find them. Some of the more simple results are listed next—but if your students are like ours, they will be able to make up plenty of their own.

$$\frac{16\!\!\!/5}{6\!\!\!/60} = \frac{15}{60}, \quad \frac{2\!\!\!/6}{\!\!\!/65} = \frac{2}{5}$$

$$\frac{38\!\!\!/5}{8\!\!\!/80} = \frac{35}{80}, \quad \frac{1\!\!\!/9}{\!\!\!/95} = \frac{1}{5}$$

$$\frac{3\!\!\!/32}{8\!\!\!/30} = \frac{32}{80}, \quad \frac{4\!\!\!/9}{\!\!\!/98} = \frac{4}{8}$$

$$\frac{27\!\!\!/5}{7\!\!\!/70} = \frac{25}{70}$$

$$\frac{26\!\!\!/66}{6\!\!\!/665} = \frac{2\!\!\!/66}{6\!\!\!/65} = \frac{2\!\!\!/6}{\!\!\!/65} = \frac{2}{5}$$

$(31 + ½)(21 − ⅓) = 31·21$
$(2 − {}^{2}/_{13})(4 + {}^{1}/_{6})(5 + {}^{1}/_{5}) = 2·4·5$

$2^5 \cdot 9^2 = 2592$
$31^2 \cdot 325 = 312325$
$73 \cdot 9 \cdot 42 = 7 \cdot 3942$
$9^2 − 1^2 = 92 − 12$
$7^2 − 3^2 = 72 − 32$

---

*The **Mathematics Teacher** (February 1971).

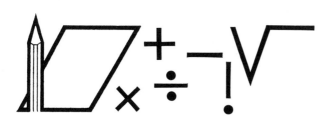

PAN-NUMERICS

$$\frac{3^4 + 5 - 6}{1} = 9^2 - 8 + 7 + 0$$

How many equations can you write, using each of the digits exactly once?

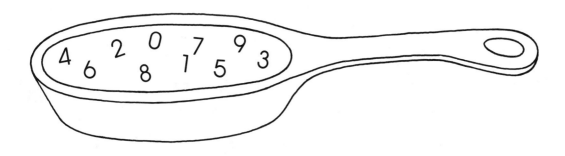

PAN-NUMERICS

This idea came from the idea of using all the letters of the alphabet in a sentence. An example is "Pack my brown box with five dozen liquor jugs." This is called a **pangram**. In mathematics, we decided to call an equation **pan-numeric** if it contained every digit **exactly once.**

If this appears too tough, we relax the requirement. We tell our youngsters to write an equation with as many digits used once as they can. We then write the equations on the board for others to see. We praise the unusual ones, the simple ones, in fact we praise for any excuse at all.

Remember, creativity, as much as any other factor, is tied to a person's attitude toward new ideas. So we try to foster in students the willingness to attempt something different—and are often surprised at what they can do.

If you study number bases other than ten, ask, "Can this problem be done in other bases?" In base two you could write $0! = 1$ (since these are the only two symbols). We can't think of any other equations in base two that fit the requirement. Maybe you can.

In base three the symbols are 0, 1, and 2—so one solution is $2^0 = 1$. Are there others?

For variation, you can give bonus points for each different operation used, or for using all operations exactly once. If you think it is appropriate, introduce the factorial notation, the greatest integer function, square roots, powers, or absolute values. These functions are defined and explained in the glossary.

 Cards Squares

CARD SQUARES

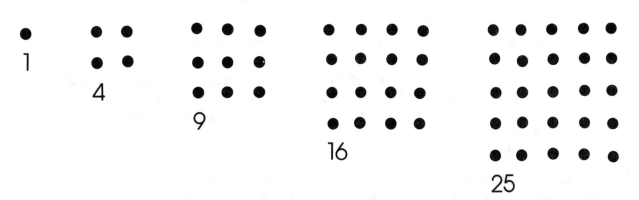

Can you finish putting the spade cards next to the heart cards so that each pair always sums to a square number?

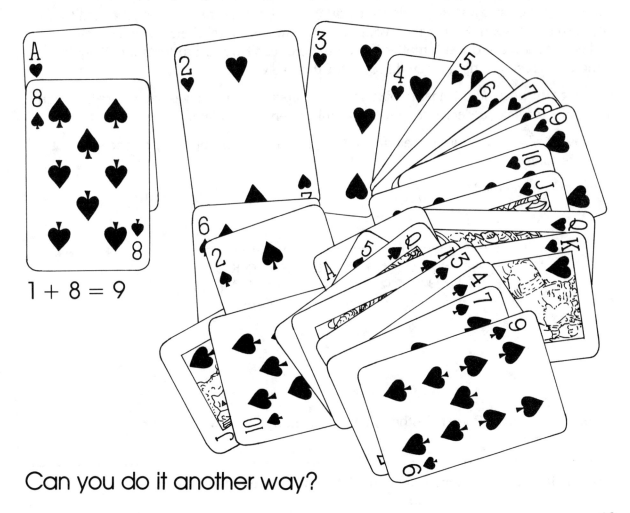

1 + 8 = 9

Can you do it another way?

CARD SQUARES

••
You will need a deck of cards for each pair of students. Construction paper facsimiles will work.
••

We got this idea from an article by Martin Gardner. He reported:

> ...David L. Silverman is the inventor of this puzzle. Remove the spades and hearts from a deck. Put the spades face up in a row in serial order with the ace at the left and the king at the right. Place a heart card under each spade so that the sum of the two cards is a square number. Prove that the solution is unique.*

After the students have solved Silverman's puzzle, you can make variations of it. For instance, will it work if you remove the kings? What if you remove both the kings and queens? If a student wants to make all face cards equal to ten, he's probably been playing the game of "21." Ask him, "How do you suppose that'll work out?" Then let him try it.

It is tempting to tell students the answer too soon, or that you've already tried something they suggest and it doesn't work so well. Resist the temptation to tell them any results too soon. Remember, your first objective is to get them to do some addition, so it really doesn't matter if the puzzle doesn't work. Students are likely to remember best the things they discover for themselves.

Eric Fromm once said "Most people fail in life because they never see where or when they have to make a decision; they **think** only when it's too late."

We say, Maybe people aren't being given a chance to **think,** but only to accept what they are told.

You can change that!

```
Red      A 2 3 4 5   6 7 8 9 10 J Q K
Black    8 2 K Q J 10 9 A 7   6 5 4 3
```

Note: Rook® cards (Parker Brothers) can be used by those who don't care to use the common type.

---

*Martin Gardner, "Mathematical Games," **Scientific American** (November 1974).

KER-PLUNK

Here are some billiard tables, all missing the same amount of bumper guard on two corners. The dotted lines show the path of a ball shot from the lower left corner.

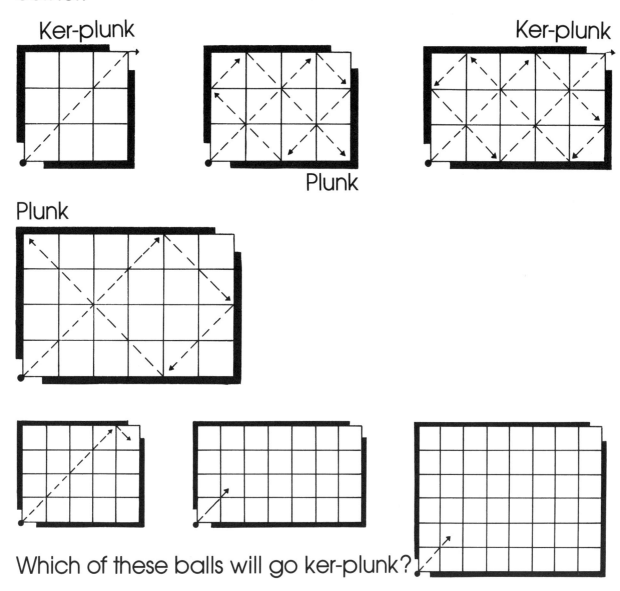

Which of these balls will go ker-plunk?

Make up different size tables and see if you can predict when the ball will go ker-plunk.

KER-PLUNK

Harold Jacobs, in his book **Mathematics a Human Endeavor***, opened the first chapter with a series of problems dealing with a billiard table and an imaginary ball that will rebound forever, unless it hits a corner. The dimensions of the table could be changed at will so that the ratio of length to width became the variable. Here is an adaptation of that theme.

Imagine a billiard table, with the lower left corner and upper right corner of the bumper guard missing, as shown in the illustrations. You are allowed to change the table's dimensions, but not the missing bumper on those corners (½ unit in both directions for the lower left and upper right corners).

Any other special conditions? Yes—you must assume the ball stops when it hits any bumpered corner (PLUNK), and that it goes off the table (KER-PLUNK) if it hits a corner with the bumper missing. Assume, also that the outside table edge will be drawn on graph paper without making new lines (so a 3½ unit table is not allowed).

The first goal is to ascertain the table dimensions necessary for the ball to finally go KER-PLUNK. Ask the youngsters to look for things that are common among those kinds of tables, and to develop a rule to predict this.

After your students have done some experimenting, encourage them to look at their data in table format—perhaps like this:

Height = h	Width = w	h/w (reduced**)	Result

Notice that on some tables the ball crosses every square. Call them **Jacobean** tables. Ask your students to look for clues as to when a table is a Jacobean table.

Here are some generalizations we got from our students:
1. Whenever h/w (reduced) is an odd number over an odd number the ball will go KER-PLUNK.
2. Whenever h/w (reduced) is such that h × w is an odd number the ball will go KER-PLUNK.
3. Whenever h/w (reduced) is such that h × w is an even number the ball will go PLUNK.
4. Jacobean tables happen whenever h/w is already reduced to begin with.

---

*W. H. Freeman and Co., San Francisco, 1970.

**Why reduced? Look at the path of a ball on a 2 × 3 table and compare it with the path on a 4 × 6, or a 10 × 15 table.

INDIANAPOLIS 500

	Jan	Feb	Mar	Apr	May	Jun	Jul	Aug	Sep	Oct	Nov	Dec
1	74	91	28	68	20	60	0	71	67	27	25	45
2	61	47	10	22	44	17	20	19	20	55	30	13
3	96	70	90	91	51	73	70	58	96	15	29	21
4	90	9	53	1	32	71	36	84	92	18	84	18
5	65	58	3	43	47	53	75	87	6	92	90	0
6	62	43	77	62	69	77	60	93	0	8	45	21
7	97	82	72	13	42	54	38	31	38	63	94	45
8	69	10	77	8	55	98	82	82	72	33	32	4
9	96	55	68	11	0	25	48	49	31	6	76	59
10	79	8	7	80	42	12	61	2	13	83	23	83
11	20	19	77	70	46	60	43	44	15	51	44	84
12	96	34	24	54	24	98	77	87	83	57	8	83
13	47	13	32	92	70	65	98	78	37	97	92	46
14	80	54	71	37	55	46	41	70	6	17	98	70
15	99	18	4	92	73	81	93	67	72	12	2	59
16	20	65	68	10	74	81	2	82	16	6	86	95
17	10	96	85	39	94	21	99	92	63	32	44	89
18	14	50	73	42	17	1	58	56	96	0	74	15
19	30	1	43	69	73	36	58	2	33	3	74	87
20	21	49	96	53	77	40	71	26	86	0	52	4
21	35	30	56	78	25	10	28	78	89	78	16	94
22	1	76	91	73	78	38	81	64	18	16	15	54
23	66	47	86	28	34	74	59	48	50	96	38	47
24	66	49	85	52	99	41	50	78	65	63	11	94
25	66	28	41	5	13	19	1	3	47	14	9	7
26	8	99	31	65	57	65	74	14	42	28	3	40
27	80	62	84	88	20	5	74	21	65	14	66	3
28	64	37	23	9	62	1	94	95	72	9	95	24
29	37	29*	89	9	13	20	40	32	29	99	1	29
30	32		42	61	76	97	35	55	19	40	6	74
31	79		60		78		16	75		99		17

*On Leap Years

INDIANAPOLIS 500

We usually tell our students the following at the beginning of this activity:

> Notice that it is possible to locate your birthday on the array of numbers. Months are designated across the top, and dates indicated down the left-hand side.
>
> Your task is to find a path of ten numbers that sum to 500. Your birthday is the first number. You may select any one of the adjacent numbers as the second number, and any number adjacent to **that** as the third number. If you play chess, you'll recognize this move from number to number as being like the king's move.
>
> After you've found a king's path with nine steps that totals 500, try some other scheme of moves: a knight's move, or, see how small or large a number you can get in ten moves.

Some of our students have become quite involved when this approach was used; alas, others were simply not interested. We then suggested that some might prefer to try a partner game in which two students mark their respective birth dates then try to find a path of ten or more, if necessary, steps between them—having a minimum or maximum total. This was successful with some students, although those previously interested did not want to change from their particular solitaire versions. All accepted versions were fine with us, since our objective was to have them do some addition problems.

PERMUPRIME PAIRS

**Exponents
Powers**

13 is prime.
31 is prime.

13 and 31 are a permuprime pair. Likewise, 103 and 031 are a permuprime pair, because 103 and 031 are both prime numbers.

But 173 and 371 are **not** a permuprime pair, because $371 = 7 \times 53$.

Problem:

Find as many permuprime pairs as you can.

Variation: Notice that 169 and 196 are both squares, $169 = (13)^2$, $196 = (14)^2$; so 169 and 196 are a permusquare pair.

Can you find other permusquare pairs? Can you find any permucube pairs?

PERMUPRIME PAIRS
EXCAVATIONS of PERMUTATIONS with
INDICATIONS
of
EXASPERATION over ALTERNATION

Permutation means "changing around." When you rearrange the furniture in your room you are doing a permutation. When you shuffle a deck of cards, you are doing a permutation. When you mix up the letters of a word, it's called an **anagram** (spot . . . tops . . . pots . . . , etc.). An anagram is a permutation of letters. Similarly, 31 is a permutation of the digits in the number 13. Because both 31 and 13 are prime, and one is a permutation of the other, we call them a **permuprime pair.**

Challenge your students to find as many permuprime pairs, or sets, as they can. One way to begin is to insert a zero between the digits of a number you know is prime, like 13, to get a new number 103. Then check to see if the new number is also prime.

Your students can use their present knowledge about prime numbers; they will likely learn more about them as they look for the permuprime sets.

Here are some results our students found:

Permuprime Sets	Permusquare Sets	Permucube Sets
11 and 11	144 and 441	125 and 512
103, 013, and 031	169, 961, and 196	
107, 017, and 071	256 and 625	
37 and 73	1024 and 2401	
113, 131, and 311	4096 and 9604	
137, 173, and 317	0225 and 2025	

Note: We got this idea from Alice Kelley, "Permutation Prime Numbers," **Journal of Recreational Mathematics,** vol. 3, no. 1 (1970).

ALPHA-SYMMETRIC

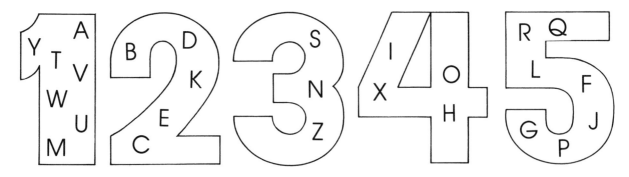

The letters in 1 have a common property. What is it?

The letters in 2 have a common property. What is it?

What is the property for 3? for 4? for 5?

If: one = 4 + 3 + 2 = 9
two = 1 + 1 + 4 = 6

o is in 4
n is in 3
e is in 2

How many entries can you find for the blanks below?

Word	Alpha-Symmetric Value	Relationship
One	9	greater than, by 8
Two	6	greater than, by 4
Three		greater than, by 11
Four		
	21	
		equal
		twice
		both prime
		both contain same digits

ALPHA-SYMMETRIC

The title of this exercise might give the students a clue as to why the alphabet characters were placed in the arrangement illustrated. If they can't figure it out, give them some hints by drawing on the blackboard the sequence of figures shown below. You might wish to draw only the first letter, with its

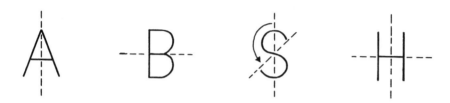

vertical axis. If necessary, after a suitable wait you can draw the second letter, and so on.

Most classes will be able to figure out that the letters

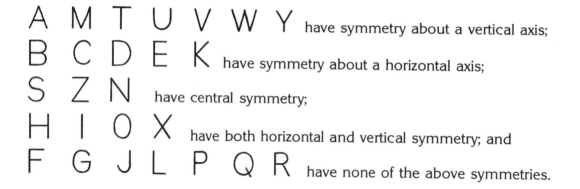

Each letter is related to a particular number; these related values of the letters in a word can be added so that an **alpha-symmetric** number is assigned to that word. For example, the word "one" has a value of $4 + 3 + 2 = 9$; so we would say that the alpha-symmetric number for the word two is obtained thus: two = $1 + 1 + 4 = 6$.

There may be many or no answers for the blanks in the table. Let your students find out which is the case. Sometimes finding that there is no answer is, in fact, the correct solution to a problem.

See George Pólya's, **Mathematics and Plausible Reasoning,** Vol. 1, (Princeton University Press, 1973), p. 89, for an interesting problem connecting this classification of the alphabet letters with equations.

TWENTIETH CENTURY PALINDROMES

June 7, 1976 is an interesting date. It can be written

6-7-76

or, ignoring the dashes, 6776. Notice the sequence of digits is the same in both directions—so it is a **palindromic date.**

How many dates can you find in the twentieth century that are palindromes when written in the form month-day-year?

Is October 1, 1901 a palindromic date?

> Can you find dates with **other** special properties?
>
> ....
> Jan. 3, 1931
> ....
>
> June 4, 1900
>
> Feb. 3, 1911

TWENTIETH CENTURY PALINDROMES
SOME MEMOS

People who work with computers frequently write dates in a special way. January 1, 1976 might be written 010176 or 1-1-76 or 1/1/76. A date written as a number, no matter how you write it, can have some interesting properties. June 7, 1976 is an example worth looking at: if you write it 6-7-76, it is a palindrome (ignore the dashes). A palindrome is an expression that reads the same from left to right as from right to left. Notice that if you wrote June 7, 1976 as 060776, it would not be a palindrome.

Here is a problem. How many dates can you find since the beginning of the twentieth century that, when written in one (or more) of these computer forms, are palindromes?

Is June 4, 1900 a palindromic date? The possible expressions are: 060400, 6-4-00, and 6/4/00. None of these would result in a palindrome although it is interesting to note that 6400 is a perfect square.

Encourage your students to look for dates that are interesting in other ways. January 3, 1931 leads to the number 1331, which is a palindrome—and even more—it is the **cube** of a palindrome as well, since $(11)^3 = 1331$.

Did you know "aibohphobia" is the fear of palindromes?

THE DIE IS CAST

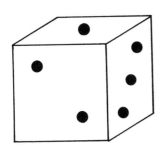

How many dots are on the side
opposite ⚀ ?
opposite ⚁ ?
opposite ⚂ ?

If and ⊠′ are opposite faces, what can you say about the number of dots in ⊠ + ⊠′ ?

Notice the pattern here:

$1+2+3= 6$ $1+3+5= 9$ $1+5+4= 10$ $1+4+2=7$
$\quad\quad\;\;6$ $\quad\quad\;\;9$ $\quad\quad\;\;\;10$ $\quad\quad\;\;7 = \boxed{\boxed{32}}$
$\quad +$ $\quad +$ $\quad\; +$ $\quad +$

Complete analogous patterns starting with these arrangements:

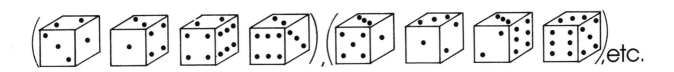

Question: If **n** is the number of dots on the top, in the patterns above, what is the expression for the number in ☐ Can you **prove** it?

THE DIE IS CAST

You'll need a model for this activity. A die, or a cube marked like a die, is required. Reproduce the pattern below so that students can make their own models. They should each have one die and become familiar with it. Eventually they may work on the problems without actually having the model at hand. If so, encourage them to sketch (literally or mentally) the various positions from memory and verify the correctness of their sketches by comparison with the model. The ability to do this is not essential for solving this particular problem, but you might explain that a mathematician often can't see the physical problem he is working on, so he must imagine it. Thus, learning how to visualize things in space is a worthwhile mental activity.

Students should observe that they can sometimes see fewer but never more than three faces at a time. They should note that the faces have one, two, three, four, five, and six spots on them and that the sum of the spots on opposite faces is always a constant—seven.

Now, using the restrictions on an ordinary die, you can pose some questions that reveal interesting number patterns. For the first question, have students place a die on their desks with the single spot on top so that the two vertical faces are visible (see the figure). Add the visible spots and record the sum. Turn the cube 90 degrees about its vertical axis. A new visible sum appears; record it. Continue until you have observed and recorded four visible sums, then add them to get a new grand total—sum of sums. Repeat the process with the two spots on top, then three, and so on.

The relationship between the number of spots on the top face, **n,** and the sum of sums generated by that face is $4n + 28$. You can argue that this is so (a proof) because you always add the top face four times, then you add each of the four vertical faces twice; but the vertical faces can be thought of as constituting four pairs of opposite faces each of whose sum is 7.

The variation is to compute, with a fixed face up, four visible **products,** then the sum of those products. The relationship between the number of spots on the top face, **n,** and the sum of products is $49n$. This can be verified by checking each of the six cases.

Which of these two dice is probably available in a neighborhood store?

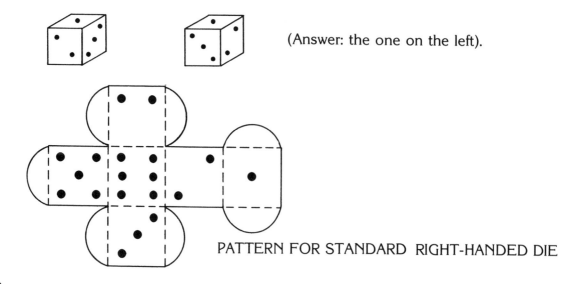

(Answer: the one on the left).

PATTERN FOR STANDARD RIGHT-HANDED DIE

Variation: suppose you change the pattern in each case to this:

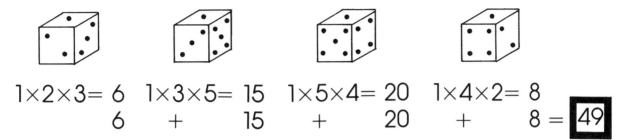

1×2×3= 6 1×3×5= 15 1×5×4= 20 1×4×2= 8
     6   +   15   +   20   +   8 = $\boxed{49}$

Super question: If **n** is the number of dots on the top, in the patterns above, what is the expression for the number in ☐? Can you **prove** it?

FAMILY HARMONY

Look at 24.

Notice:

```
     12              6                    4
2 ) 24          4 ) 24        2 + 4 = 6 ) 24
    2              24                   24
    ‾              ‾‾                   ‾‾
    4               0                    0
    4
    ‾
    0
```

Are there other numbers such that each digit divides the number **and** the sum of the digits divides the number?

Consider 288.

```
    144                36                    16
2 ) 288           8 ) 288    2 + 8 + 8 = 18 ) 288
    2                 24                    18
    ‾                 ‾‾                   ‾‾‾
    8                 48                   108
    8                 48                   108
    ‾                 ‾‾                   ‾‾‾
    8                  0                     0
    8
    ‾
    .0
```

:OK here: :OK here: :OK here:

It works!

How many more numbers like this can you find?

FAMILY HARMONY

Let's say 24 is the family name of a group of numbers. The members of this family are 2, 4, and 6 (2 and 4 are the digits in the number, and 6 is the sum of the digits). Observe the harmonious relationships, in this case, between the individual members and their family name. The member 2 divides 24 evenly, as do 4 and 6. So we say this family is harmonious.

Are there other harmonious families? If so, how many and how can we find them?

Begin by experimenting with a few numbers you know well. Consider, for example, the family name 288 (perhaps that's too gross). Some of its members are 2, 8, and 8. Since 8 and 8 are twins they should, of course, be treated equally—both should be included in all family activities—so the other important family member is the sum of the numbers, $2 + 8 + 8 = 18$. As you can see from the illustration, 288 is also harmonious.

In fact, many numbers are harmonious—see how many you and your students can find.

GOLDBACH'S* CONJECTURE

Every even number greater than 4 is the sum of two odd prime numbers.

thus   6 = 3 + 3 ·····one way only

8 = 3 + 5 ·····one way only

but

10 = 5 + 5

and   10 = 3 + 7 ·····two ways!

Can you find an even number than can be written as the sum of two primes in three ways? how about four ways? or more?

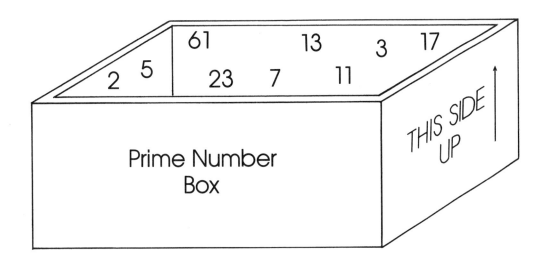

*Pronounced: gold-box-con-**ject**-your. When you say "It seems, on the basis of some data, that all _____ are _____ , but I can't prove it," you are making a **conjecture**.

GOLDBACH'S CONJECTURE

Goldbach's conjecture was first made by Christian Goldbach (1690–1764) in a letter to the Swiss mathematician Leonhard Euler. What is most interesting about Goldbach is that although he was a capable and respected mathematician of his day, it was not the problems he solved that earned him a place in history, rather, the one he could **not** solve. He knew how to ask the right question: "Can every even number greater than 4 be expressed as the sum of two primes?" All the evidence indicated to Goldbach that the answer was yes, but no one has ever **proved** it.

Ordinary people accept the truths of others; special people ask only to be allowed to make their own definitions of truth. In this search, it is often more important to ask the right questions than to have the right answer to someone else's question.

We have asked some other questions about Goldbach's conjecture. We're fairly sure that the answers would be hard to find and prove, in general; but we're also confident that the pursuit of the answers will provide students practice with prime numbers and addition. So have your students try them—who knows, someone may yet prove Goldbach's conjecture.

6 = 3 + 3	8 = 3 + 5	10 = 5 + 5	10 = 3 + 7
12 = 5 + 7	14 = 7 + 7	14 = 3 + 11	16 = 5 + 11
16 = 3 + 13	18 = 7 + 11	18 = 5 + 13	20 = 7 + 13
20 = 3 + 17	22 = 11 + 11	22 = 5 + 17	22 = 3 + 19
24 = 11 + 13	24 = 7 + 17	24 = 5 + 19	26 = 13 + 13
26 = 7 + 19	26 = 3 + 23	28 = 11 + 17	28 = 5 + 23
30 = 13 + 17	30 = 11 + 19	30 = 7 + 23	32 = 13 + 19
32 = 3 + 29	34 = 17 + 17	34 = 11 + 23	34 = 5 + 29
34 = 3 + 31	36 = 17 + 19	36 = 13 + 23	36 = 7 + 29
36 = 5 + 31	38 = 19 + 19	38 = 7 + 31	40 = 17 + 23
40 = 11 + 29	40 = 3 + 37	42 = 19 + 23	42 = 13 + 29
42 = 11 + 31	42 = 5 + 37	44 = 13 + 31	44 = 7 + 37
44 = 3 + 41	46 = 23 + 23	46 = 17 + 29	46 = 5 + 41
46 = 3 + 43	48 = 19 + 29	48 = 17 + 31	48 = 11 + 37
48 = 7 + 41	48 = 5 + 43	50 = 19 + 31	50 = 13 + 37
50 = 7 + 43	50 = 3 + 47	52 = 23 + 29	52 = 11 + 41
52 = 5 + 47	54 = 23 + 31	54 = 17 + 37	54 = 13 + 41
54 = 11 + 43	54 = 7 + 47	56 = 19 + 37	56 = 13 + 43
56 = 3 + 53	58 = 29 + 29	58 = 17 + 41	58 = 11 + 47
58 = 5 + 53	60 = 29 + 31	60 = 23 + 37	60 = 19 + 41
60 = 17 + 43	60 = 13 + 47	60 = 7 + 53	62 = 31 + 31
62 = 19 + 43	62 = 3 + 59	64 = 23 + 41	64 = 17 + 47
64 = 11 + 53	64 = 5 + 59	64 = 3 + 61	66 = 29 + 37
66 = 23 + 43	66 = 19 + 47	66 = 13 + 53	66 = 7 + 59
66 = 5 + 61	68 = 31 + 37	68 = 7 + 61	70 = 29 + 41
70 = 23 + 47	70 = 17 + 53	70 = 11 + 59	70 = 3 + 67
72 = 31 + 41	72 = 29 + 43	72 = 19 + 53	72 = 13 + 59
72 = 11 + 61	72 = 5 + 67	74 = 37 + 37	74 = 31 + 43
74 = 13 + 61	74 = 7 + 67	74 = 3 + 71	76 = 29 + 47
76 = 23 + 53	76 = 17 + 59	76 = 5 + 71	76 = 3 + 73
78 = 37 + 41	78 = 31 + 47	78 = 19 + 59	78 = 17 + 61
78 = 11 + 67	78 = 7 + 71	78 = 5 + 73	80 = 37 + 43
80 = 19 + 61	80 = 13 + 67	80 = 7 + 73	82 = 41 + 41
82 = 29 + 53	82 = 23 + 59	82 = 11 + 71	82 = 3 + 79
84 = 41 + 43	84 = 37 + 47	84 = 31 + 53	84 = 23 + 61
84 = 17 + 67	84 = 13 + 71	84 = 11 + 73	84 = 5 + 79
86 = 43 + 43	86 = 19 + 67	86 = 13 + 73	86 = 7 + 79
86 = 3 + 83	88 = 41 + 47	88 = 29 + 59	88 = 17 + 71
88 = 5 + 83	90 = 43 + 47	90 = 37 + 53	90 = 31 + 59
90 = 29 + 61	90 = 23 + 67	90 = 19 + 71	90 = 17 + 73
90 = 11 + 79	90 = 7 + 83	92 = 31 + 61	92 = 19 + 73
92 = 13 + 79	92 = 3 + 89	94 = 47 + 47	94 = 41 + 53
94 = 23 + 71	94 = 11 + 83	94 = 5 + 89	96 = 43 + 53
96 = 37 + 59	96 = 29 + 67	96 = 23 + 73	96 = 17 + 79
96 = 13 + 83	96 = 7 + 89	98 = 37 + 61	98 = 31 + 67
98 = 19 + 79	100 = 47 + 53	100 = 41 + 59	100 = 29 + 71
100 = 17 + 83	100 = 11 + 89	100 = 3 + 97	

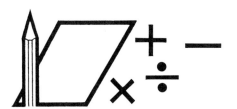

Statement:

Whenever you reverse the digits in a number and subtract the smaller number from the larger, the answer is a multiple of 9.

Do you **believe** it?
Can you **prove** it?

$$\begin{array}{r} 864 \\ -468 \\ \hline 396 \end{array}$$

Question:

Suppose you reverse the digits in a number and multiply the two numbers. When will the answer be a multiple of 9?

$12 \times 21 = 252^* = 9 \times 28$
$13 \times 31 = 403$, not a multiple of 9
$15 \times 51 = 765 = 9 \times 85$

Look at a lot of examples and see if you can make up a rule that will tell you when a number multiplied by its reverse will be a multiple of 9.

---

*A palindrome too!

×LASREVER×

Imagination is a mathematician's most valuable ally. If you can imagine something that doesn't exist, you may be on your way to becoming a mathematician. Imagine a magic wand that can convert any number into another number that is a multiple of nine. There is just such a magic wand. Do this: Pick any number you like, with two or more digits, reverse the order of the digits, and find the difference between the two numbers. The difference will **always** be a multiple of nine. It will even work with numbers less than ten if you place one or more zeros in front of them (2 is 02, 4 is 0004, etc.)

Ask your students to see if this particular kind of magic will always work, then ask if anyone can explain why. Most "magic" is explainable **if** you search deeply enough. This reminds us of another valuable friend of the mathematician, curiosity—the attribute that leads us in search of the **reasons** for the way things happen.

In this case, the reason can be seen by first assuming t and u are digits with t larger than u. Then a two-digit number may be represented as
$$10t + u,$$
with reversal $10u + t$, which is smaller than the original number.
Then subtracting, $(10t + u) - (10u + t) = 9t - 9u = 9(t - u)$, which is a multiple of 9.

In case $u = t$ we get the special case where $(t-u) = 0$, which is nine times zero. Finally, if u is larger than t the required difference would be $9(u-t)$, again, a multiple of nine.

Similar demonstrations can be given for numbers with 3, 4, 5, ... digits.

The second part of this section involves a nagging question. If you take the product of a number and its reversal, under what circumstances will the answer be a multiple of nine? Your students should eventually see that it is true whenever the sum of the digits in the original number is a multiple of three. Let them discover this.

Some students prefer questions that ask "Why is this so? Others prefer the ones that ask "Under which conditions is this true?"

When students finish this activity, they may be aware that whenever the sum of the digits in a number is a multiple of 9 the number is then divisible by 9. If no student discovers this, you may want to have them fill in the following table.

Number, n	Sum of digits	Remainder after dividing n by 9
1	1	1
2	2	2
.	.	.
.	.	.
.	.	.
22	4	4
23	5	5
.	.	.
.	.	.

FACTOR POLYGONS

24 factors into $2 \times 2 \times 2 \times 3$.

Notice:

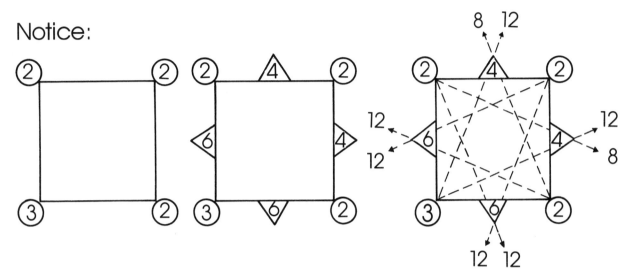

All these numbers are proper divisors of 24.

....

Can you make similar diagrams for any number that is the product of three or more primes?

Will this always give you **all** the number's proper divisors?

....

Try $60 = 2 \cdot 2 \cdot 3 \cdot 5$

Try $30 = 2 \cdot 3 \cdot 5$

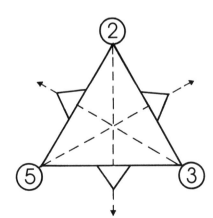
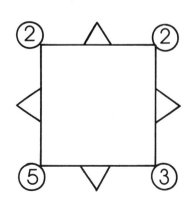

FACTOR POLYGONS

Mr. Larry Corina showed us these polygons. He discovered that if you first factor a number into its prime factors, then assign the prime factors to the vertices of a polygon you can make interesting diagrams. Of course the polygon has to have the same number of sides as there are prime factors of the original number.

Here is how the method works. Consider, for example, the number 24, with prime factors 2, 2, 2, and 3. Since it has four prime factors, put each of them at one corner of a four-sided polygon (see the first illustration; this happens to be a square but any four-sided convex polygon would do). Now, in the middle of each side of the polygon draw a small triangle and write the product of the two adjacent corner numbers in each triangle. Next, draw in the dotted lines from each triangle to the non-adjacent corners and write the products of the numbers you find on those dotted lines. In the case of 24, all of the numbers you can see are factors (or **divisors**, or **aliquot parts**) of 24. And there are no other factors of 24 except for 24. So this polygon shows all of 24's proper divisors.

This method will always give you some, but not always **all,** of the divisors of a number. Encourage your students to find out when it will show all the proper divisors and when it won't, and **why.**

It is helpful to look at an example in which it doesn't work. Consider the number 72, which factors into 2 × 2 × 2 × 3 × 3.

One resulting factor polygon is shown here (you can draw others if you place the prime factors in different order around the polygon).

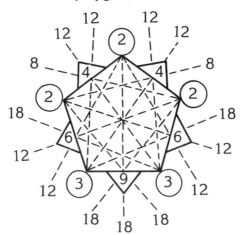

Notice that 24, 36, and 72 don't appear. These are the aliquot parts of 72 with more than three factors. If you check where the numbers come from, this is no surprise.

So, whenever the original number has more than four prime factors this method won't show all the aliquot parts of the number. Nevertheless, your students will find many cases in which it works and may be able to figure out variations that will give them **all** the factors.

You might suggest that they first guess how many of the numbers from 12 to 144 it will work for, then check to see if their guesses are correct.

+LASREVER+

73 is 37 backwards.

```
   73
+  37
  110
```
is divisible by 11.

1472 is 2741 backwards.

```
  1472
 +2741
  4213
```
is divisible by 11.

....

            Is this always so?
            Does it work for six-digit numbers?
            How about three-digit numbers?

....

Can you find a rule to tell which numbers it will work for? Can you **prove** it?

    How to think of two-digit numbers? Perhaps, 10**t**+**u**, where **t** and **u** are digits.

+LASREVER+

This exercise will give your students an opportunity to discover that whenever you add a number with an even number of digits to that number with its digits reversed, you get a number that is a multiple of 11. Let them experiment with various numbers until they guess that it is true. Then, you may wish to demonstrate it more formally.

Begin by calling **u** the units digit, **t** the tens digit, **h** the hundreds digit, **T** the thousands digit, and so on. Thus, any two-digit number might be represented by
$$10t + u,$$
and reversed, $10u + t$.
Adding the two numbers yields $11t + 11u = 11(t+u)$, which is divisible by 11.

Similar proofs can be given for all numbers with an **even** number of digits. But what about an odd number of digits? Consider, as an example, any three-digit number represented by
$$100h + 10t + u$$
and reversed, $100u + 10t + h$.
Adding the two gives $101u + 101h + 20t = 101(u+h) + 20t$, which is not always divisible by 11. However, in special cases it might be. Thus, if **u, h** and **t** were chosen to be **2, 2,** and **4,** respectively, it would be divisible by 11.

If your students are able to follow this much of the discussion, suggest that they see how many three or five digit numbers have this special property.

KAPREKAR'S CONSTANT

Choose a four-digit number in which not all digits are alike. Write the digits in descending order, then in ascending order and subtract. Watch!

Choose: 4823

      8432 descending
     −2348 —ascending
      6084

Repeat the process with 6084.

      8640 descending
     −0468 —ascending
      8172

Keep repeating until you **know** you should stop!

> Can you make up a process, something like this one, so that nice things happen?

KAPREKAR'S CONSTANT

This topic gives students a chance to use their creativity. First present it as shown on the illustrated page. Put an example on the blackboard, but let each student choose his own four-digit number. Let the students discover that when you obtain 4176 the next number will be 6174 and the next 6147, and so on. So there is no point in going on. Then suggest that they try to create some other process to see if anything interesting happens. You will want to show them some examples. Here are some from our students:

1. Choose any four-digit number (not all digits alike). Write it first in descending order, then in ascending order and add. Repeat the process with the last four digits of the answer.
2. Choose any three-digit number. Multiply by 3. Keep the last three digits of the answer and repeat the process. (This can be varied by choosing a different size number, or by multiplying by other numbers.)

If some students need practice with multiplication, suggest that they try the second example—but have them multiply by a small number at first. When they report to you about some pattern they've noticed, congratulate them for being so perceptive and ask, "Do you think it will work for the next bigger number?" When you reward them for noticing some pattern, they will do quite a bit more computation. And there is always the chance that a student will reward you by discovering a truly remarkable pattern.

DIGITADITION SEQUENCES

14, 19, 29, 40, 44, 52, 59, ___?___ , . . .

What comes next?

___?___ , 68, 82, 92, 103, 107, . . .

What came just before 68?

If this sequence is constructed like the ones above can there be any number before 20?

_____ , 20, 22, 26, 34, 46, 56, . . .

Answer: No. Find out why not.

20 is called a **self number.** How many self numbers can you find?

....

Are there any numbers with two (or more) antecedents?

DIGITADITION SEQUENCES

D. R. Kaprekar, a mathematician from Devali, India, invented this kind of number sequence. Let your students discover this kind of number sequence. If they have trouble, show them part or all of the following array:

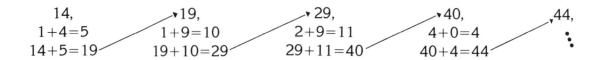

Then show them another that works the same way, say,

18, 27, 36, 45, 54 ____ . ······ 63, 72, 81, 90, 99, 117

Then, another one built on the same principle,

56, 67, 80, 88, 104, 109, 119, 130, 134, ____ . ······ 142

When they have figured out the method for finding the next term, have them try to find the term that comes **before** each beginning term (the antecedent). For example, in the sequences above, 7 comes before 14, 9 before 18, 46 before 56. Finding the antecedent for a number is not always possible and it is not always unique when it exists. No number comes before 20—try all the possibilities and you'll see why. If a number has no antecedent it is called a **self number.** Many self numbers exist, so your students will have quite a bit of success in finding them. There are also some numbers with more than one antecedent. It is a little like digging in the sand—every once in a while you find something nice. (And it promotes considerable addition practice.) After your students have constructed some sequences of their own, tell them about the following number trick.

···

Kaprekar says: To find the sum of all the digits in any one of the digitadition sequences all you have to do is (1) subtract the first term from the last term, and (2) add the sum of the digits of the last term to that result.
···

So, according to Kaprekar's method, the sum of the digits in:

14, 19, 29, 40, 44, 52, 59, 73 is
     73 − 14 + 7 + 3 = 69.

Check the result to verify that it is really equal to:

1 + 4 + 1 + 9 + 2 + 9 + 4 + 0 + 4 + 4 + 5 + 2 + 5 + 9 + 7 + 3.

Martin Gardner reported that Kaprekar said, "The proof of all this rule is very easy and I have completely written it with me. But as soon as the proof is seen the charm of the whole process is lost, and so I do not wish to give it just now."* Suggest that your students try it and verify it enough to decide whether to believe Kaprekar. Some of them may be able to prove it.

See if you are as surprised by the responses as we were when we asked our students some of these questions:

1. Do you believe Kaprekar's statement: Why?
2. Can you determine if it will always work? Is that a proof?
3. What is the difference between believing something and proving something?
4. Can you name some things that you believe, but you cannot prove?
5. Can you name some things that you can prove, but you don't believe?

---

*Martin Gardner, **Scientific American** (March 1975)

CRITTENDEN'S LETTER

Here is a picture of a letter sent to us by R. W. Crittenden:

Look at it carefully — what is special about the stamps? What is special about 3211123?

Suppose you had 1, 2, 3, 4, 5, 6, and 8 cent stamps and envelopes like this:

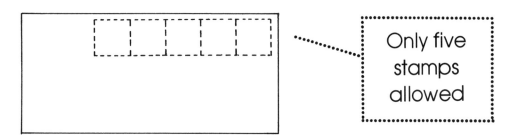

Only five stamps allowed

If you must put 13¢ postage on each of your letters, how many "interesting" numbers can you create?

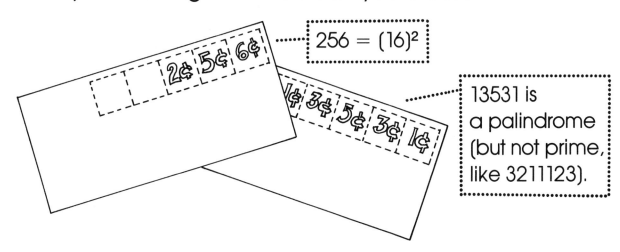

$256 = (16)^2$

13531 is a palindrome (but not prime, like 3211123).

CRITTENDEN'S LETTER

If sending a letter cost a penny, and only one cent stamps were printed, sending a letter wouldn't be particularly interesting to a mathematician. But in 1976 the cost for a first class letter in the United States was 13 cents. Stamps were available in many denominations: 1, 2, 3, 4, 5, 6 and 8 cent stamps. The size of our envelopes created an additional constraint—they were too small for 13 one-cent stamps.

Ask your students to assume an envelope only has space for 5 stamps and suggest some of the following questions:

1. How many different ways could you put those kinds of stamps on an envelope with places for just five stamps?
2. What are all the possible values of postage you can get?
3. If you were a spy and wanted to send coded messages, how many different messages could you send?
4. Say that each different arrangement is a number represented by the value of the stamp in that particular position, then make a table of the numbers and see how many interesting facts about each of those numbers you can find. Mr. Crittenden, for example, is interested in palindromes, so he was happy to find the arrangement in the picture. He went further and discovered that there are exactly thirteen prime palindrome arrangements for 13¢ postage. Are you interested in any kinds of numbers? See if your interest is represented by the postage and stamp problem.

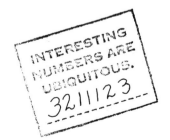

Professor Jean Pedersen
Department of Mathematics
University of Santa Clara
Santa Clara, CA 95053

SPECIAL AREAS*

Has an area of 4 square units and a perimeter of 8 units.

* Has an area of 16 square units and a perimeter of 16 units.

* Has an area of 22 square units and a perimeter of 22 units.

....

1. How many other shapes can you find such that their area and perimeter have the same numerical value?

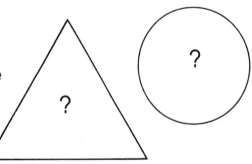

2. How many shapes can you find such that their area is half their perimeter?

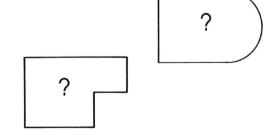

Copyright © 1979 by Addison-Wesley Publishing Company, Inc.

55

SPECIAL AREAS

Give your students some graph paper and present this idea as shown in the illustration. Perhaps you'll need to explain that area is the number of squares inside the boundary and perimeter is the length, in units, around the boundary. Allow them plenty of time to experiment with the problem on their own before giving any more clues or suggestions.

If you have previously discussed areas of right triangles, circles, or other special figures you might extend this idea to these figures.

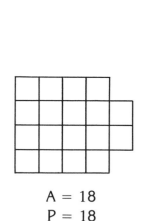
A = 18
P = 18

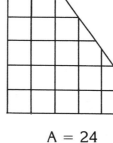
A = 24
P = 24

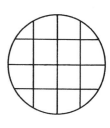
A = 4π
P = 4π

Don't present this too soon, though, let them think of it themselves if they can.

You can ask students these questions (our students responded well):
1. Can you find any different figures (like a triangle, circle, or combinations of those) with the perimeter equal to the area?
2. Guess how many such figures are possible.
3. Do you believe your guess?
4. Can you prove your guess?
5. Can you think of a situation where you, or the teacher, have proven something but, you don't really believe it?

We have found that the answers to the last question frequently enlighten us about our students and our teaching.

SPECIAL VOLUMES
One of nature's secrets

Ten cubes can be arranged like this:

Volume = 10
Surface Area = 36

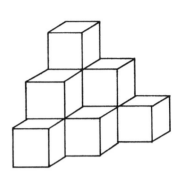

Ten cubes can also be arranged like this:

Volume = 10
Surface Area = 34

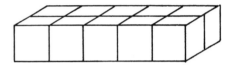

How small can you make the surface area by rearranging the 10 cubes?

Use some cubes to find the numbers that go in this table:

Volume	Maximum Surface Area	Minimum Surface Area
10	42	30
11		
12		
13		
14		
15		
.		
.		
.		
125		

SPECIAL VOLUMES
ONE OF NATURE'S SECRETS

This topic has far-reaching implications in nature that you may want to point out to your students by means of the questions below.

Begin by showing a cube, and explain that people sometimes say, "Its volume is one and its surface area is six." More precisely, what they really mean is that the area of each of the six square faces comprising its surface is one **square** unit and the volume of the cube is one **cubic** unit. It is customary, however, in ordinary conversation to drop the specific name of the unit when confusion is not possible. Specifying a particular unit of measure would not affect the results of this problem, so let us consider the relationship between the numbers representing volume and surface area in some kind of unspecified unit. For example, if we construct a figure that looks like three cubes side by side, we get 3 and 14 for the volume and surface, respectively.

Let your students experiment to find the relationship between volume and surface area for given stacks of unit cubes. Have them begin with 10 cubes and organize a separate table of data for the results—finally recording maximum and minimum numbers. Then have them find the minimum and maximum surface areas for 11, 12, . . . cubic units.

The following questions will relate this exercise with the role of volume and surface area in nature:

1. What seems to be the general shape that has the least area for a given volume?
2. Does this have anything to do with the shape of soap bubbles?
3. Does this have anything to do with building a house? a submarine?
4. Some biologists say the cooling capacity of animals is directly related to their surface area. Does this help to explain why desert animals, like snakes, are long and thin? Do the shapes of other desert animals confirm this explanation?
5. Does this tell us anything about what gives whales the ability to tolerate temperature extremes?
6. Does it help to explain why furry animals curl up in a ball to hibernate?
7. How do biologists measure surface area and volume for various animals? First guess how they do it, then verify your guess, if you can.

RAPID CALCULATIONS

$15^2 = \underline{2}25$ $\quad\quad$ $\underline{2} = 1 \times 2$
$25^2 = \underline{6}25$ $\quad\quad$ $\underline{6} = 2 \times 3$
$35^2 = \underline{12}25$ $\quad\quad$ $\underline{12} = 3 \times 4$
$45^2 = \underline{20}25$ $\quad\quad$ $\underline{20} = 4 \times 5$
$\vdots$
$105^2 = \underline{110}25$ $\quad\quad$ $\underline{110} = 10 \times 11$

Can you see a pattern?

If you have found a pattern can you explain it? can you prove it?

Here are some number patterns:

$0 \times 9 + 2 = 11$ $\quad\quad$ $1 \times 8 + 1 = 9$
$1 \times 9 + 2 = 11$ $\quad\quad$ $12 \times 8 + 2 = 98$
$12 \times 9 + 3 = 111$ $\quad\quad$ $123 \times 8 + 3 = 987$
$123 \times 9 + 4 = 1111$ $\quad\quad$ $1234 \times 8 + 4 = 9876$
$1234 \times 9 + 5 = 11111$ $\quad\quad$ $12345 \times 8 + 5 = 98765$
$\cdots$ $\quad\quad\quad\quad\quad\quad\quad\quad\quad$ $\cdots$

$0 \times 9 + 8 = 8$
$9 \times 9 + 7 = 88$
$98 \times 9 + 6 = 888$
$987 \times 9 + 5 = 8888$
$9876 \times 9 + 4 = 88888$
$\cdots$

How far does each of these patterns go before a contradiction is reached, if ever? Why?

What similar patterns can you find?

RAPID CALCULATIONS

These number patterns and the one displayed below were shown to us by Jim Salvi, a high school teacher in San Jose, California.

The first pattern on the student page involves squaring a number ending in five. The rule (or pattern) might be described as: Multiply the integer preceding the 5 by the next higher integer and attach 25.

Proof: Express the number as $10a + 5$, where $a = 1, 2, 3, 4, \ldots$, not necessarily less than 10.

Then, $(10a + 5)^2 = 100a^2 + 100a + 25$
$= 100a(a + 1) + 25.$

The other number patterns will eventually reach a contradiction, as can be verified by actual computation, but students who investigate them will learn some properties of numbers and multiplication. Perhaps your students will be able to find some interesting patterns of their own.

....

$1^2 = 1$
$11^2 = 121$
$111^2 = 12321$
$1111^2 = 1234321$
· · ·
How far?

$91^2 = 8281, 2 = 1 \times 2, 81 = (10-1)^2$
$92^2 = 8464, 4 = 2 \times 2, 64 = (10-2)^2$
$93^2 = 8649, 6 = 3 \times 2, 49 = (10-3)^2$
· · ·
How far?

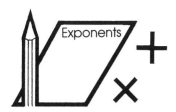

POWER

Rule 1. Each term in these sequences must have an exponent greater than the exponent of the previous term.

Rule 2. To obtain the text term you must either multiply two preceding terms or square one of the preceding terms.

Example: Here is such a sequence, where a^{15} is obtained from a in six steps.
$a, a^2, a^4, a^8, a^{12}, a^{14}, a^{15}, \ldots$

Possible entries for the next term are:
$a^{16}, a^{18}, a^{19}, a^{20}, a^{22}, a^{23}, a^{24}, a^{26}, a^{27}, a^{28}, a^{29}, a^{30}$.

Do you see how to get each of them?

Question:

Could you get a^{15} in fewer than six steps?

....

What is the least number of steps in which you can obtain $a^{20}, a^{25}, a^{30}, a^{35}, \ldots$ etc.?

Start with a and obey rules 1 and 2.

POWER

This is the kind of problem encountered by computer scientists. Of course computers can rapidly compute a^{15} in six steps, but suppose the computation had to be carried out several thousand times. Then it would be better to do it in five (or fewer, if possible) steps—computer time is expensive.

Let your students make a chart on the board and keep a record of the shortest ways to arrive at various powers of **a.** First, ask for the minimal sequence to a^{15}—then to a^{20}, a^{25}, a^{30}, etc. You can consider a^n for any **n** if you wish.

Notice that a^{15}, for example, can be obtained in **five** steps by at least two routes:
$a, a^2, a^3, a^5, a^{10}, a^{15}$
or $a, a^2, a^4, a^5, a^{10}, a^{15}$

We don't believe a^{15} can be obtained, using rules 1 and 2, in fewer than five steps. However, if any of your students find a way, let us know—write in care of the publisher.

> Recall
> $\underbrace{aaaaa \ldots a}_{n \text{ factors}} = a^n$, so $aaaa = a^4$
> $a^n a^m = a^{n+m}$, so $a^2 a^3 = a^{2+3} = a^5$
> $(a^n)^2 = a^n a^n = a^{n+n} = a^{2n}$, so $(a^5)^2 = a^5 a^5 = a^{5+5} = a^{10}$
> $a^0 = 1$, because $a^0 \cdot a^1 = a^{0+1} = a^1$.

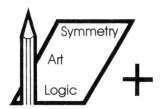

ALBRECHT DÜRER

Challenge: Write each of the numbers
   1, 2, 3, 4, 5, 6, 7, 8, 9, 10, 11, 12, 13, 14, 15, 16
in a square of this array so that the sums in all rows, columns, and diagonals are the same.

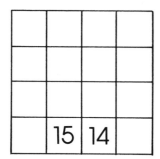

In honor of Albrecht Dürer, who included this particular magic square in his now-famous engraving, **Melancholia,** in the year 1514, start with the numbers 15 and 14 as shown.

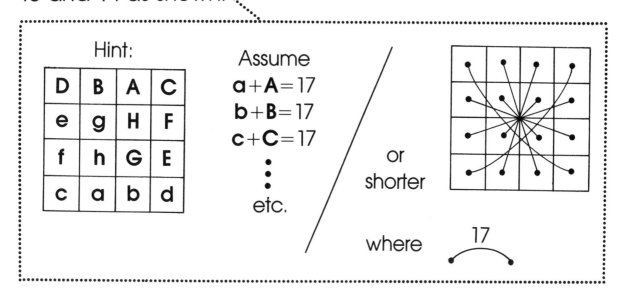

Read about Albrecht Dürer to find out some of the ways mathematics is related to art.

ALBRECHT DÜRER

The solution of this particular magic square is a pretty exercise in logic (details below). When you encourage your students to investigate Albrecht Dürer they may discover how he applied projective geometry to perspective drawing. His techniques are still used by artists today.

A SOLUTION:

Begin by determining **what** the sum in each row, column, and diagonal should be. This is done by first computing $1+2+3+4+5+6+7+8+9+10+11+12+13+14+15+16 = 136$, then reasoning that this sum must be evenly divided among the four rows (or columns). So, the required sum in any row (or column) must be $\frac{136}{4} = 34$.

Now, by the hint, we know that **A** = 2 and **B** = 3. The bottom row tells us that $c + 15 + 14 + d = 34$, or $c + d = 5$.

The possibilities for **c** and **d** may be seen in the equations:
$1 + 4 = 5$,
$2 + 3 = 5$,
$3 + 2 = 5$,
$4 + 1 = 5$.

Only the first and last equations are possibilities, since we have already used the numbers 2 and 3. Choose **c** = 1 and **d** = 4 (suggest that your students verify that the reverse choice would also work).

Now, using the hint again, we find that **C** = 16 and **D** = 13. Next, consider **e** and **f**. They must satisfy $13 + e + f + 1 = 34$, or $e + f = 20$. Remember that the only numbers now available are 5, 6, 7, 8, 9, 10, 11, and 12. Thus, the only choices for **e** and **f** are given by the equations:
$e + f = 12 + 8$,
$e + f = 8 + 12$,
$e + f = 11 + 9$,
$e + f = 9 + 11$.

Suppose we choose **e** = 12 and **f** = 8 (again, other choices will work, as may be verified by your students). This means, by the hint, that **E** = 5 and **F** = 9.

To determine the values of **g, G, h, H**, we have only to appropriately assign the numbers 6, 7, 10, and 11. Continuing as before, we can now make choices for **g** and **h** so that $g + h = 16$, and $g + G = 17$, $h + H = 17$, and the sum in each row and column is still 34. There is a small margin for error in terms of the choices at this point, but the number of possible choices is small, so success is likely.

The completed solution looks like this:

13	3	2	16
12	6	7	9
8	10	11	15
1	15	14	4

At certain times during the solution, choices between two equally plausible possibilities were necessary. Ask your students to try solving Dürer's square by making **other** choices to see how many different magic squares they can find.

Suppose, then, that you don't require the 15 and 14 to be in any specified squares—and that you can use the hints implied by the diagrams below—how many magic squares can you create?

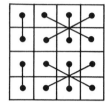 , 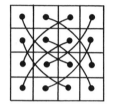 , 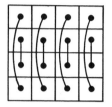 , where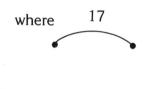

Here are some possible solutions; exchanging appropriate rows and columns will give many others.

16	3	10	5
1	12	7	14
8	13	2	11
9	6	15	4

1	14	7	12
15	4	9	6
10	5	16	3
8	11	2	13

1	14	12	7
4	15	9	6
13	2	8	11
16	3	5	10

Melancholia

DIGITAL ROOTS

Assertion:

7 is called the digital root of 214, because 2 + 1 + 4 = 7.

214 ÷ 9 will have a remainder of 7.

Is this true? Why?

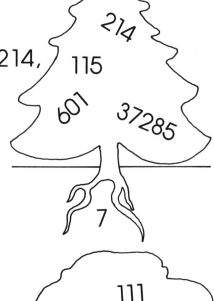

Assertion:

7 is called the digital root of 37285, since
3 + 7 + 2 + 8 + 5 = 25
and 2 + 5 = 7.

So 37285 ÷ 9 will have a remainder of 7.

Is this true? Why?

....

4762109 ÷ 9 will have a remainder of what?

Guess: 4 + 7 + 6 + 2 + 1 + 0 + 9 = 29
(calculated) 2 + 9 = 11
  1 + 1 = **2** is remainder

DIGITAL ROOTS

The repeated addition of the digits of a number until a single-digit answer is obtained is called finding the **digital root** of the original number. This is used in the process of "casting out nines," which may be studied after this section.

Since the digital root is a single digit, by definition, the expected outcomes are 0, 1, 2, 3, 4, 5, 6, 7, 8, 9. The digit 0 occurs only when you start with the number zero. The digit 9 occurs whenever a number is a multiple of 9. Digital roots other than 0 and 9 always indicate the remainder that results when the original number is divided by 9.

Here are some hints to help your students understand why the digital root always represents the remainder you get after dividing the original number by 9. First,
$$214 = 2(100) + 1(10) + 4$$
$$= 2(99 + 1) + 1(9 + 1) + 4$$
$$= 2(99) + ②+ 1(9) +①+④.$$
Notice that all terms except the circled ones are evenly divisible by 9 so a remainder of $2 + 1 + 4$ results.

Similarly, for 37285 we write,
$$37285 = 3(10000) + 7(1000) + 2(100) + 8(10) + 5$$
$$= 3(9999 + 1) + 7(999 + 1) + 2(99 + 1) + 8(9 + 1) + 5$$
$$= 3(9999) +③+ 7(999) +⑦+ 2(99) +②+ 8(9) +⑧+⑤.$$
Again, all terms except the circles ones are evenly divisible by 9, so a remainder of $3 + 7 + 2 + 8 + 5 = 25$ results. Then, since 25 is larger than 9, we repeat the process writing:
$$25 = 2(10) + 5$$
$$= 2(9 + 1) + 5$$
$$= 2(9) +②+⑤.$$
All terms except $2 + 5 = 7$ are divisible by 9 so 7 is evidently the **proper** remainder.

SHORT CUTS

Notice in the number 37285 the 7 and 2 add to 9. So, you can toss out those and simply compute the digital root as:
$$3 + 8 + 5 = 16$$
$$1 + 6 = 7.$$
To see why this works, write the number in expanded form as in the examples above and see what happens to the numbers you tossed out. Notice that although the numbers obtained in intermediate steps may vary, the final answer for the digital root must be the same.

Students will gain most from this topic by choosing their own numbers and finding the digital roots by various means and checking them by division. Working many examples will convince most students, but some demonstrations by expanding the numbers and using the property that $10^n = \underbrace{9999\ldots9}_{n \text{ digits}} + 1$, as above, will help to reinforce students' understanding of some of the worthwhile properties of the number base system.

Check:
```
                    529123
                9 ┌─────────
                  │ 4762109
                    45
                    ──
                     26
                     18
                     ──
                      82
                      81
                      ──
                       11
                        9
                       ──
                        20
                        18
                        ──
                         29
                         27
                         ──
                          2  = R. OK.
```

> Choose some numbers of your own and see if this always works. See if you can either find a number for which it won't work, or show why it must always work.

GRANDFATHER'S MODELS
Do Digital Roots First

Grandfather checks his computations this way:
DR(**n**) means the digital root of **n**.

```
  2345} 5 ⎤
  9742} 4 ⎥ 5 + 4 + 3 + 4 = 16
  5025} 3 ⎥              1 + 6 = 7
+ 1948} 4 ⎦
+ 18780} 6 ················ Wrong
```

He is thinking ·······
> DR(2345) = 5, so 2345 = 9**a** + 5
> DR(9742) = 4, so 9472 = 9**b** + 4
> DR(5025) = 3, so 5025 = 9**c** + 3
> DR(1948) = 4, so 1948 = 9**d** + 4
> The digital root of the answer should
> = DR(5+4+3+4) = 7. But, DR(18780) = 6
> and 6 ≠ 7 so **the computation is wrong!**

Here is another one:

```
   2345} 5 ⎤ 5 × 4 = 20
 × 9472} 4 ⎦       2 + 0 = 2
22211480} 2  ··················· The answer might be correct.
```

Questions: Why doesn't Grandfather say the answer is correct?

When will this method verify that you have a computational error?

Can you explain why it works?

GRANDFATHER'S MODELS

Aircraft designers always make a model of their designs and test them in a wind tunnel. The Wright Brothers used this idea. If the model doesn't test well the full-size airplane won't fly well. But the converse isn't true: if the model flies well in the test, the full-size airplane may, or may not, fly as well as the model.

Consider that. Think about situations in which evidence in one case is always true, but the reverse situation may or may not be true. For example, Mrs. Jones' car is in the parking lot at a supermarket, therefore it is not at home in the garage. However, if Mrs. Jones' car is not in the garage, it is not necessarily at the supermarket.

With that idea in mind, consider Grandfather's models for checking computations. He can model an arithmetic problem and find that if the **model** problem is solved correctly and gives proper results, the actual problem may or may not be computed correctly; but if the model shows a different result than the actual problem, the latter is certain to be in error. Further, the model problem is so much easier than the actual problem that he can almost always solve the model in his head.

The basis for constructing the model involves a concept called digital roots. Recall that adding all the digits in a number, then adding them again, repeatedly, will finally leave a single digit (the remainder you would get by dividing the original number by 9)—the digital root. The interesting thing is this: If you solve a problem with digital roots, and take the digital root of the answer, that will be the digital root of the answer in the full size problem.

To understand why this is the case, consider the second example:

```
         2345
       × 9472
       ——————
       22211840
```

Digital root of 2345

$DR(2345) = 5$, so $2345 = 9a + 5$
$DR(9472) = 4$, so $9472 = 9b + 4$.
thus, $2345 \times 9472 = (9a + 5)(9b + 4)$
$= (9a)(9b) + (9a)(4) + (5)(9b) + (5)(4)$.

But each of the first three terms has 9 as a factor so,
$DR(2345 \times 9472) = DR(5 \times 4) = DR(20) = 2$.

If 22211840 is the correct answer, then $DR(2211840)$ must equal 2. Indeed, it does, so there is a strong inference that the problem is correct. But we can't be absolutely certain since it is also true that

$DR(22211804) = 2$,
$DR(22211840) = 2$,
$DR(82221140) = 2$, etc.

and any one of those might be the correct answer on the same grounds. (You might rule out the first one if you realize that the last digit must be 0.) However, there is only one correct answer—so we can conclude that when Grandfather's model works, it shows only that the computation given might be correct; or, if it is erroneous, that it differs from the correct answer by a multiple of nine.

SHORT CUTS

Consider this computation, which was checked by Grandfather's model:

$$
\begin{array}{r}
23\cancel{45} \} \cancel{5} \\
\cancel{9}4\cancel{72} \} \cancel{9} \\
5025 \} 3 \\
+ \cancel{1}94\cancel{8} \} 4 \\
\hline
\cancel{1}87\cancel{9}0 \} 7
\end{array}
\Bigg\} \; 3 + 4 = 7
$$

May be ok

Notice the crossed-out digits. That is part of a quick method to arrive at the digital root of any number, called **casting out nines.** In the above example, we arrived at the 5 on the right by casting out (throwing away) the 5 and 4 of the first addend (because they equal 9), leaving 3 and 2, which equal 5. In the second addend, we cast out the 9 and also the 7 plus the 2, leaving just the 4 as the digital root. We couldn't cast out any nines from the third addend, so we added the two 5s and the 2 to get 12, then added the 1 and the 2 of 12 to get 3—the digital root of that addend. For the last addend, we cast out the 9, and the 8 plus the 1, leaving the digital root of 4. Now, look at the right-hand column of figures. We can cast out the 5 and 4 and arrive at the sum of 7, which will always be the digital root of the correct answer.

Demonstrate this method to your students and suggest that they try it both ways with some numbers to verify it.

NO DUPLICATES

461 × 146 = 67306 ⋯⋯ Two 6's

421 × 142 = 59782 ⋯⋯ No duplication of digits

85 × 58 = 4930 ⋯⋯ No duplication of digits

Challenge:

Find all the numbers you can such that the product of the original number and the number that results from moving the last digit to the first position has no duplication of digits.

Variation: Add or subtract the original and rearranged numbers, trying to get answers with no duplications.

NO DUPLICATES

Ask each student to write their favorite three-digit numbers. Then instruct them to: Move the digit on the right to the left end of the number, in order to form a new number. Next find the product of the two numbers. Does the product have any repeated digits? If not, they were lucky, because that is the kind of product desired in this exercise.

Have your students figure out how many different computations are necessary to investigate all two-digit numbers. (Hint: When you start with 29 and get the product 29×29, you have also computed the product you would get if you started with 92. Fifty-five computations are required if you consider 00, 01, 02, . . . 09 as two-digit numbers.) Ask students to guess how many of the two-digit numbers will yield no duplications when treated this way. Record their guesses on the board before they do the computations.

Now consider all three-digit numbers. How many computations will have to be made to examine all cases? (When you start with 123 you get 123×312, but the computation isn't the same for 312—that product is 312×231. Consequently, nearly one thousand computations are required to examine all three-digit numbers. Note that 001, 002, etc. are considered three-digit numbers.)

Since there are so many computations to consider for three-digit numbers, it may be helpful to have the students guess how many of the one thousand computations would yield no duplicates. Record their guesses on the board, then break the class into groups and let each group check various three-digit numbers, for example, those from 000 to 199, or 200 to 399. The students in each group may wish to subdivide the problem.

As a special challenge, who can find the largest number that, when treated this way, yields no duplicates.

•••

For a variation, you can do something similar with the sums or differences of the numbers. If you, or your students, thought of that before reading this page, congratulations! You're becoming interested in a subject called **number theory.** Guess what variations might be suggested in the next exercise you look at.

MENTAL LIGHTNING

Problem:
    48
× 16
   ?

Step 1.
    48
× 16
   ··8

    Think, 6 × 8 = 48
    Write ··8

Step 2.
    48
× 16
 ··6 ··8

    Think, (6 × 4) + (1 × 8) + $\overbrace{4}^{\text{dots}}$ = 36
    Write ··6

Step 3.
    48
× 16
7 ··6 ··8

    Think, (1 × 4) + $\overbrace{3}^{\text{dots}}$ = 7
    Write 7

Practice until you can write the answer without hesitation for two-digit multiplications.

See if you can figure out the sequence of steps for three-digit times two-digit computations.

MENTAL LIGHTNING

If you practice this method enough to become proficient at it, let your students give you any two different two-digit numbers and just write the product on the board—astounding one and all!*

Since our students never quite believe we're capable of such feats without some special knowledge we are forced to show them the steps 1, 2, and 3. It is important to explain just why it works.

In step 1 you obtain the units digit by multiplying units by units—writing down the units digit and indicating the tens by dots. In step 2 you get the tens digit by taking units by tens in the two possible ways and adding the dots from step 1—writing down the tens digit and indicating the hundreds by dots. Finally, in step 3 you obtain the hundreds digit by multiplying the tens digits together and adding the dots from step 2. If you get a two-digit number in step 3, write both digits down. The whole process is sometimes abbreviated graphically as follows:

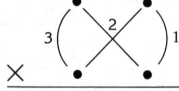

In this scheme, three-digit numbers multiplied by two-digit numbers would look like this:

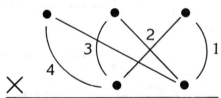

    1 represents multiplications for the units position
    2 represents multiplications for the tens position
    3 represents multiplications for the hundreds position
    4 represents multiplications for the thousands position

This activity has been successful with both high ability and underachieving students. Apparently being able to **immediately** write answers to problems that require some pencil and paper work provides a great deal of satisfaction for most students. As you know, students will work quite hard to master the method if they are given a reward—such as a chance to be the "star," or for many, just your praise.

We saw this idea in **Mathematics for Elementary Teachers** (Wadsworth Publishing Co., Belmont, Ca.) by Alice J. Kelly and David E. Logothetti.

---

*Don't get the answers right every time. As any magician will testify, an occasional error creates excitement.

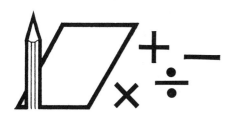

COLUMBUS DAY DISCOVERY

1. ~~1~~ ~~2~~ 3 ~~4~~ 5 6 7 8 ~~9~~

	1	
4	9	2

Can you put the rest of these numbers in the blank squares—so that the sum of every row, column and diagonal is always the same? This is called a **magic square.**

2.

	6	

Can you complete this magic square so that you use nine consecutive numbers?

....

3. Can you construct a magic square with 7 in the center? How about 8? 9?

4. Can you construct a magic square with nine consecutive even numbers?

5. Can you construct a magic square with nine different fractions?

COLUMBUS DAY DISCOVERY

When you introduce this, tell the students you already have an answer but want them to find one so you can check your answer. Emphasize that the sum in each of the directions shown in (a) should be the same. They should then deduce, from the bottom row, that the desired sum is 15. Let the students work out their solutions. They will most likely arrive at the solution shown in (b). You can then write your solution, (c), on the board. You might ask them if they see any advantage in writing the solution your way.

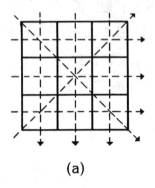

8	1	6
3	5	7
4	9	2

(5+3)	(5−4)	(5+1)
(5−2)	5	(5+2)
(5−1)	(5+4)	(5−3)

(a)   (b)   (c)

Leave the two solutions on the board and ask if they can find the answers to any of the other questions. Notice that if all nine numbers are known, you can compute the sum in any given direction by adding all the numbers and dividing by 3. Thus in 1, the sum in each direction should be

$$\frac{1+2+3+4+5+6+7+8+9}{3} = \frac{45}{3} = 15.$$

There is then a nice way to determine which number should be in the center cell. It can be illustrated as follows:

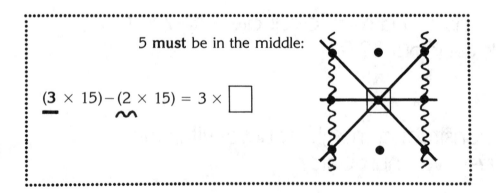

5 **must** be in the middle:

(3 × 15) − (2 × 15) = 3 × ☐

You may wish to show this to your students and let them try to figure out what it means. [Hints: The sum of the numbers on any line (indicated ▬▬ or ∿∿) should be 15. The center number, □, appears on the **three** lines indicated by ▬▬.]

If you know 15 should be the sum in each direction, there is another way to determine the relative positions of the digits. Observe what happens when you write 15 as the sum of three digits, in ascending order, from the set 1, 2, 3, 4, 5, 6, 7, 8, 9; in all possible ways. There are exactly eight ways to do this and those correspond with the sums in the 3 rows, 3 columns, and 2 diagonals. Thus

$$\begin{align}15 &= 1 + 5 + 9 \\ &= 1 + 6 + 8 \\ &= 2 + 4 + 9 \\ &= 2 + 5 + 8 \\ &= 2 + 6 + 7 \\ &= 3 + 4 + 8 \\ &= 3 + 5 + 7 \\ &= 4 + 5 + 6.\end{align}$$

Since 5 appears **four** times it must be in the center square. Likewise, 2, 4, 6, and 8 appear **three** times each, so those must be placed in corner squares; while 1, 3, 7, and 9 each appear exactly twice, so those must be placed in the remaining edge squares.

Articles about magic squares can be found in various issues of **The Mathematics Teacher** and **The Arithmetic Teacher** (consult the index in the December issue for specific articles). There is also a book on the subject, **Magic Squares and Cubes,** by W. S. Andrews (Dover Publications, 1960). If your students express interest in this area they will be able to gain much from this book.

COMMUNITY PROPERTY

Each of these has the property			None of these has the property			Which of these has the property?	
A 828 252 8055			249 137 2123			470	1669
720 6021 963			6364 4677 1897			270	8550
873 576 810			6538 325 140			482	408
24282 963 738			618 20431 159			315	504
297 9612 49608			578 9092				
8262 522			219 280333				
B 29665		81921	35560		39127	65526	17200
39986		81586	64321		32956	71522	74201
17454		74665	67931		20648	28627	23072
93853		95610	27016		34959	22841	
91103		35411	32697		95053	92854	
80760		74683	10552		210162		
77340		92811	37985				
C 59452		38432	29663		97589	17039	59231
28032		76372	38346		34129	12816	89928
86222		43210	65961		84356	48703	69232
47090		45132	86714		37705	73433	91370
17304		90560	93785		88397	14212	36827
34242		21633	86837		82716	53946	67579
D 1267 1299 1778			7491		2019	7987	1551
4467 3689 1246			2868		9670	0238	1377
2336 1567 3467			4417		7167	1445	1126
7899 6779 2999			3013		2545	8054	7046
5699 2234			2060		1920	8042	4484
3555 6689			2725		9151	1779	4556

COMMUNITY PROPERTY

You might introduce this exercise by giving your students a warm-up problem. Tell them to look at this community of numbers:
   2   828   6   522   49608   8262   10   14.
Do they have any property in common? Yes, they are all even numbers.

Then have them look at this community of numbers:
   3   297   5   873   137   46809   15677   83
Do they have any property in common? Yes, they are **not** even numbers; they are all odd numbers.

Have them next look at this set of numbers, and ask, "Which numbers have the same property as the first community of numbers?"
   1   4   325   2124   618   8055   578   579

The answer is: 4, 2124, 618, 578.

You **could** arrive at the correct answer without stating the property of odd or even. All that is required in these puzzles is to decide whether each number in the third set has the property possessed by all the members in the first set. After the first set is given, several hypotheses are possible. The second set of numbers may elimate some of the hypotheses, as long as it does not eliminate all of them, you can find an answer.

The purpose of this exercise is to get students to use and test inductive reasoning. We set up the problems so that the distinguishing characteristics involve arithmetic, that is, i.e., addition, multiplication, subtraction, and division. We have avoided giving more information on the problem page than what is absolutely necessary. Since this may be too difficult for some students, here are some hints to be used only when they would otherwise give up. Give one hint at a time, or better, start a discussion such as: If you were making up problems like this, what properties would you choose? Would you select odd and even? (Probably not, because it is too obvious.)

If this inspires your students to discuss the properties of numbers, all these hints will probably come up; if not, use the list:
    1. Is there anything interesting about the sum of the digits (see **A, B,** and **C**)?*
    2. Is there anything interesting about the product of the digits?
    3. Is there anything interesting about the shape of the symbols?
    4. Is there anything interesting about the number before or after the number given?
    5. Is there anything interesting about the factors fo the number?
    6. If the number were converted to another form, would ther by anything interesting about it?
    7. If you sequenced the numbers in some way, would the sequence have missing terms? Would the missing numbers be interesting?

Hint for **D:** $7 \leq 9 \nleq 8 \nleq 7$, no. $\;0 \leq 2 \leq 3 \leq 8$, yes.

---

*Interesting could mean: odd, even, prime, composite, multiples of certain numbers, powers of some other number, digits made from straight lines, sums of pairs of digits, square numbers, triangular numbers, Fibonacci numbers, digits in ascending or descending order, palindromic, etc.

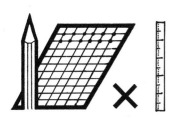

DOTS

Start with a 5 unit × 5 unit square arrangement of dots, like this:

Notice that you can connect pairs of dots to form 4 straight line segments whose intersection forms a square,

like this: or this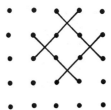

A 3×3 square area, 9 square units

Can't tell length of side but area looks like 2 square units

Questions:

1. How many **different** size squares can you make this way?

2. How many of those squares have an area represented by a **whole** number of square units?

3. What is the **smallest** size square you can make this way?

DOTS

If your students have geoboards, they can be used for this exercise. If not, graph paper is almost as good, since they will need to record their answers on paper anyway.

The idea is to let the students experiment to find squares of different sizes by connecting the dots on a 5 unit by 5 unit array. They will not usually think of the third and fourth cases below—or others like them. We prefer **not** to give students the hint, in the hope that some one in class will think of it. We then congratulate them for their discovery, making a real production out of it, handshakes, ribbons, certificates of praise, inscribing names on monuments (a section of the blackboard), etc.

If you have taught the Pythagorean theorem, it will be useful in this exercise. Here are some squares our students found:

Area = 8 sq. units   **Area** = 10 sq. units   **Area** = ?   **Area** = 4½ sq. units

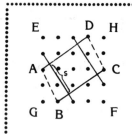

Area of $\triangle ADE$ = Area of $\triangle CBF$ = 3 sq. units
Area of $\triangle ABG$ = Area of $\triangle CDH$ = 1 sq. unit
Area of $\square ABCD$ = 16 − 3 − 3 − 1 − 1 = 8 = BC · s
But $2^2 + 3^2 = (BC)^2$, so $BC = \sqrt{13}$
Thus $8 = \sqrt{13} \cdot s$ or $s = 8/\sqrt{13}$
Consequently, **Area** = $64/13 = 4^{12}/_{13} = 4.923$

VARIATIONS

Use a 6 unit by 6 unit array of dots.
Ask for the perimeter of the square as well as the area.

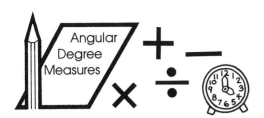

CLOCKS

The angle between the clock hands at 1:00 is 30° (thirty degrees).

The angle between the clock hands at 2:00 is 60° (sixty degrees).

What is the angle between the clock hands at 9:00?

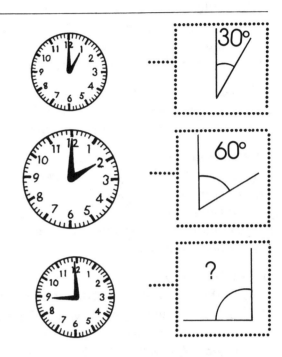

Complete this table:

Time	Angle Between Clock Hands
12:00	0°
1:00	30°
2:00	60°
3:00	90°
4:00	
5:00	
6:00	
7:00	
8:00	
9:00	
10:00	
11:00	

Questions:

1. At what other times (approximately) will the angle between the clock's hands be 90°?

2. Can you figure out the exact times?

Can you answer the above questions for 30°? 60°?

Copyright © 1979 by Addison-Wesley Publishing Company, Inc.

CLOCKS

In the early 1970's some people wore small clocks on their wrists. They called them wristwatches. Clever Boy Scouts could use a wristwatch as a compass. It was also possible to use one of these old-fashioned clocks as a goniometer, (go-nee-AH-mah-ter), a device for measuring angles.

Science has progressed: Today many people use digital time pieces—both for large clocks and wristwatches. This may have some advantages, but it is hard to imagine just how one might use a digital wristwatch as a compass or a goniometer.

How do you use a clock as a goniometer? Your eyes may be imprecise at measuring angles directly, but they can be used on an old-fashioned clock, to measure degrees with reasonable accuracy of certain angles. This is possible because your eyes are quite perceptive in determining when two lines are **parallel,** even if one of those lines is the minute hand of a wristwatch. Using this principle, we present an exercise involving clock hands and angles.

Look at the student page and cross your eyes slightly so that the image of the straight lines in the box appear to coincide with the hands on the corresponding clock. If you can do this you'll be convinced that those two angles are the same size. You may not be able to do it (one of the authors can always do it easily and the other only when well rested). You will probably feel certain that the angles are approximately equal anyway, because their respective sides **appear** to be parallel.

SOLUTIONS TO QUESTIONS 1 AND 2

Observe that the minute hand revolves through 30° in five minutes or through 6° in one minute; simultaneously, the hour hand revolves through 30° in 60 minutes, or through ½° in one minute. Thus, the minute hand revolves 5½° per minute faster than the hour hand. We have already shown that the angle between the hands at 9:00 is 90°, so start there. The minute hand will always have to gain a total of 180° on the hour hand between times when the hands of the clock will form a 90° angle. Dividing 180° by 5½° gives a time span of 32-8/11 minutes. Consequently, the next time the hands will form a 90° angle is at 9:32-8/11 (or at 9:32.727272 . . . = 9:32.$\overline{72}$; the bar over 72 means those digits are repeated indefinitely). Complete results appear in the table on page 90.

To find the times when the angle between the hands is 30°, proceed in the same way. First note that the clock hands form an angle of 30° at 1:00. In this case, the minute hand will have to gain a total of 60° on the hour hand before the hands next form a 30° angle. Dividing 60° by 5½° gives the time span of 10-10/11 minutes. So, at 1:10-10/11, or 1:10.$\overline{90}$, the hands form an angle of 30°. The minute hand will now have to gain 300° on the hour hand in order for the hands to next form a 30° angle.

(To see why this is so, turn the hands of a clock and look at the times when angles of 30° are formed.) Dividing 300° by 5½° gives a time span of 54-6/11 minutes. Now you must add 54-6/11 minutes to the time 1:10-10/11. Do this with the numbers in fractional form, remembering that when the number of minutes exceeds 60, change to the next hour (clockwise) and reduce the number of minutes by 60. The computation might look like this:

$$\begin{array}{r} 1:10\text{-}10/11 \\ +\ :54\text{-}6/11 \\ \hline 1:64\text{-}16/11 \end{array} = 1:65\text{-}5/11 = 2:05\text{-}5/11 = 2:05.\overline{45}$$

The minute hand will continue to make angles of 30° with the hour hand as it gains alternately 60° and 300° compared with the hour hand. These angular measures correspond with time spans of 10-10/11 minutes and 54-6/11 minutes, respectively. So you generate the numbers shown in the table by starting at 1:00 and adding alternately 10-10/11 and 54-6/11, until you arrive at 1:00 again.

A similar approach is used to determine when the hands form an angle of 60°. The results are given on the table.

The angle measures given in the examples and on the chart were chosen between 0° and 180°. This is the smallest degree measure for each of those angles, but it is not a unique designation and might reasonably be some other number under suitable interpretations. For example, if you imagine the minute hand sweeping through 30° every 5 minutes, then it is sensible to talk about the minute hand revolving through 270° in 45 minutes, or through 720° in two hours, or 900° in two and a half hours. It would not be wrong, then, to consider the angle between the clock hands at 7:00 as 210°. Likewise, the angle between the hands at 1:00 could be thought of as 330°. This is a useful concept in engineering and science. If you want to explain this to your students, use questions like these:

Through how many degrees does the minute hand revolve:
1. between 12:15 p.m. and 1:20 P.M.? (answer: 390°)
2. in a 24 hour day? (answer: 8640°)
3. between consecutive times when the angle between the minute hand and the hour hand is 90°? (answer: 196-4/11°)
4. between consecutive times when the angle between the minute hand and the hour hand is 180°? (answer: 392-8/11°)
5. between consecutive times when the angle between the minute hand and the hour hand is 0°? (answer: 392-8/11°)

Time Span, in Minutes, Between Occurences	10-10/11 and 54-6/11	21-9/11 and 43-7/11	32-8/11 and 32-8/11
Desired Angle Between Clock Hands	30°	60°	90°
Times	1:00 = 1:00.$\overline{00}$ 1:10-10/11 = 1:10.$\overline{90}$ 2:05-5/11 = 2:05.$\overline{45}$ 2:16-4/11 = 2:16.$\overline{36}$ 3:10-10/11 = 3:10.$\overline{90}$ 3:21-9/11 = 3:21.$\overline{81}$ 4:16-4/11 = 4:16.$\overline{36}$ 4:27-3/11 = 4:27.$\overline{27}$ 5:21-9/11 = 5:21.$\overline{81}$ 5:32-8/11 = 5:32.$\overline{72}$ 6:27-3/11 = 6:27.$\overline{27}$ 6:38-2/11 = 6:38.$\overline{18}$ 7:32-8/11 = 7:32.$\overline{72}$ 7:43-7/11 = 7:43.$\overline{63}$ 8:38-2/11 = 8:38.$\overline{18}$ 8:49-1/11 = 8:49.$\overline{09}$ 9:43-7/11 = 9:43.$\overline{63}$ 9:54-6/11 = 9:54.$\overline{54}$ 10:49-1/11 = 10:49.$\overline{09}$ 11:00 = 11:00.$\overline{00}$ 11:54-6/11 = 11:54.$\overline{54}$ 12:05-5/11 = 12:05.$\overline{45}$ 1:00 = 1:00.$\overline{00}$	2:00 = 2:00.$\overline{00}$ 2:21-9/11 = 2:21.$\overline{81}$ 3:05-5/11 = 3:05.$\overline{45}$ 3:27-3/11 = 3:27.$\overline{27}$ 4:10-10/11 = 4:10.$\overline{90}$ 4:32-8/11 = 4:32.$\overline{72}$ 5:16-4/11 = 5:16.$\overline{36}$ 5:38-2/11 = 5:38.$\overline{18}$ 6:21-9/11 = 6:21.$\overline{81}$ 6:43-7/11 = 6:43.$\overline{63}$ 7:27-3/11 = 7:27.$\overline{27}$ 7:49-1/11 = 7:49.$\overline{09}$ 8:32-8/11 = 8:32.$\overline{72}$ 8:54-6/11 = 8:54.$\overline{54}$ 9:38-2/11 = 9:38.$\overline{18}$ 10:00 = 10:00.$\overline{00}$ 10:43-7/11 = 10:43.$\overline{63}$ 11:05-5/11 = 11:05.$\overline{45}$ 11:49-1/11 = 11:49.$\overline{09}$ 12:10-10/11 = 12:10.$\overline{90}$ 12:54-6/11 = 12:54.$\overline{54}$ 1:16-4/11 = 1:16.$\overline{36}$ 2:00 = 2:00.$\overline{00}$	9:00 = 9:00.$\overline{00}$ 9:32-8/11 = 9:32.$\overline{72}$ 10:05-5/11 = 10:05.$\overline{45}$ 10:38-2/11 = 10:38.$\overline{18}$ 11:10-10/11 = 11:10.$\overline{90}$ 11:43-7/11 = 11:43.$\overline{63}$ 12:16-4/11 = 12:16.$\overline{36}$ 12:49-1/11 = 12:49.$\overline{09}$ 1:21-9/11 = 1:21.$\overline{81}$ 1:54-6/11 = 1:54.$\overline{54}$ 2:27-3/11 = 2:27.$\overline{27}$ 3:00 = 3:00.$\overline{00}$ 3:32-8/11 = 3:32.$\overline{72}$ 4:05-5/11 = 4:05.$\overline{45}$ 4:38-2/11 = 4:38.$\overline{18}$ 5:10-10/11 = 5:10.$\overline{90}$ 5:43-7/11 = 5:43.$\overline{63}$ 6:16-4/11 = 6:16.$\overline{36}$ 6:49-1/11 = 6:49.$\overline{09}$ 7:21-9/11 = 7:21.$\overline{81}$ 7:54-6/11 = 7:54.$\overline{54}$ 8:27-3/11 = 8:27.$\overline{27}$ 9:00 = 9:00.$\overline{00}$

DOMINO ARITHMETIC

Look at these two dominoes:

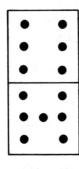
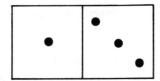

Look at the arithmetic problem—can you see the similarity?

....

Now, a problem: Using a standard double-nines set of dominoes, construct as many sets, like this, as you can; then count the **remaining** (unused) dominoes. Our best score to date is 25. Can you do it so that there are fewer than 25 dominoes left over when you're done?

DOMINO ARITHMETIC

A set of double-n's dominoes is a set of pairs of numbers, unordered, that represents all possible combinations of the numbers 0, 1, 2, . . . n. So the smallest possible set of dominoes would only have one domino—the double-blank. The next smallest set would have three—a double-blank, a blank-one, and a double-one. The next set would have a 0-0, 0-1, 0-2, 1-1, 1-2, and 2-2; for a total of six members. The next six sets of dominoes would have 10, 15, 21, 28, 36, 45, and 55 members, respectively. The double-nines set, used for this exercise, consists of fifty-five different dominoes.

Any domino (also called a **card**) can be considered a pair of digits. The one-two card, for example, could be looked at as a one above a two (1/2), a twelve (12), a two above a one (2/1), or a twenty-one (21). In the case of the illustration on the students' page, the seven-six card represents an addition problem—seven plus six. And the sum is represented by the card one-three, placed so as to represent thirteen.

After using the seven-six and the one-three, we are left with 53 dominoes. The problem is to use as many as possible of those remaining to form different addition problems. You can readily see that you can't make a problem totaling thirty-one or thirteen with the remaining dominoes because you've already used the one-three in the previous problem.

There is a solution for this problem. Consider the highest possible sum—nine plus nine. This will use the 9-9 and the 1-8 cards. Then use all the remaining nine cards except the 9-0.

You have now used **all** the cards possessing a one; in other words, you have used all the teens and the tens. At this point you have used 18 dominoes and have 37 remaining. All you have left with which to make sums are blanks. You can't use the 0-0, you have already used the 1-0, so eight usable blanks remain. You can't use the 2-0 or 3-0 because you've already used the 1-1 and 1-2. Now you have only six possible pairs. Use the 5-4 or the 7-2 or the 6-3 with the 9-0. Use the 6-2, the 5-3, or the 4-4 with the 8-0. Use the 5-2 or the 4-3 with the 7-0. Use the 3-3 or the 4-2 with the 6-0, the 3-2 with the 5-0. And finally, use the 2-2 with the 4-0. No more possibilities exist. We've used six more pairs for a total of 12 plus the 18 we used in the first group to make a grand total of 30 used and 25 remaining.

This solution, proving that 25 is the minimum, was discovered by Jennifer Pedersen (the author's daughter) when she was in the seventh grade.

If your students enjoy this problem and do well, you might suggest some of the following exercises as a continuation:
1. Use the dominoes as before, but make multiplication problems.
2. Use two dominoes to represent a fraction and its decimal equivalent. Be careful to state clearly where in the array you intend the decimal to occur.

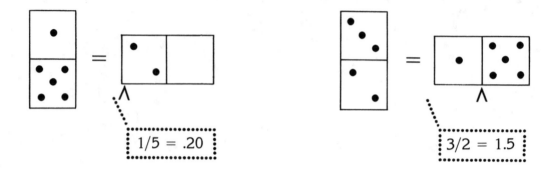

3. Use two dominoes for a total of four numbers. Find the sum.
4. Do any of the foregoing problem ideas with a different size set of dominoes.
5. Calculate how many dominoes would be in a set of double 12s? double 15s? double-ns. (See page 98.)

ODD SUMS

Look at the following:

```
1                     = 1
1 + 3                 = 4
1 + 3 + 5             = 9
1 + 3 + 5 + 7         = (   )
1 + 3 + 5 + 7 + 9     = (   )
1 + 3 + 5 + 7 + 9 + 11 = (   )
```

Do you see a pattern?

Can you **prove** it?

 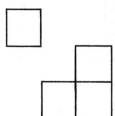 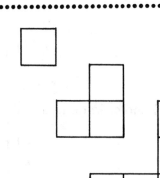

ODD SUMS

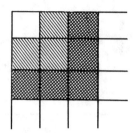

In the exercise on dominoes, we talked about the size of a given set of dominoes. A double-blank set of dominoes has one member, a double-ones set has three, a double-twos, six, and so on. We did that hoping it would arouse curiosity about what pattern would enable you to predict the size of a set if you were given the largest digit. If your students did discover a pattern, it may have been this: A set of dominoes is as large as the sum of all consecutive integers, beginning with one and ending with a number that is one larger than the largest number in the set. That is, a double-sixes set has 1 + 2 + 3 + 4 + 5 + 6 + 7 dominoes. A double-nines set has 1 + 2 + 3 + 4 + 5 + 6 + 7 + 8 + 9 + 10 dominoes.

This exercise is also concerned with the sum of a set of integers, namely, the sum of all consecutive odd integers beginning with 1. We hope that the students first discover that the sum of all consecutive odd integers is always a square number.

One way to demonstrate that this is reasonable is to show the diagram illustrated above. Begin counting in the upper left-hand corner, with the single unshaded square. In the shaded area (▨) next to that are 3 squares, and this completes an area of 4 (or 2^2) squares. In the crosshatched (▓) area are 5 squares; the total now makes an area of 9 (or 3^2). Consequently, if you continue to add the next odd number of squares to any given square you will always create a square whose side is one unit larger than the side of the last square.

Thus, $1 + 3 + 5 + \ldots + (2n - 1) = n^2$ (the sum of the first **n** odd numbers is equal to **n** squared).

We think it is exciting to be able to demonstrate a result like this with a geometric diagram. This kind of diagram is sometimes referred to as a **behold** proof.

Here is another, of the same type:

Look:

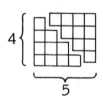

$$1+2+3+4 = \frac{4(5)}{2}$$

And, **behold,** in general:

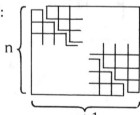

$$1+2+3+4+\ldots + n = \frac{n(n+1)}{2}$$

Notice that this is precisely the formula needed to compute the number of dominoes in a double-**n** set.

GLOSSARY

Absolute Value of a real number Its value without regard to sign; its numerical value. The number 3 is the absolute value of both -3 and 3. This is designated, in symbols, by $|-3| = 3$ and $|3| = 3$. Likewise, $|0| = 0$.

Cube Number A number that is the cube of a positive integer, such as 1, 8, 27, 64 ...

Cube Root The cube root of a number **n**, denoted $\sqrt[3]{n}$, is the number **a**, such that $a \times a \times a = n$.
Examples: $\sqrt[3]{8} = 2$
$\sqrt[3]{-27} = -3$

Even Number A number that is exactly divisible by 2 with no remainder. All even numbers can be written in the form 2n, where **n** is an integer. Thus, 2, 4, 6, 8, ... are positive even numbers. Likewise $-2, -4, ...$ are negative even numbers.

Factorial The product of all the positive integers less than or equal to the integer. Factorial n is usually denoted by the symbol n! Thus, for example,
$1! = 1$
$2! = 1 \times 2 = 2$
$3! = 1 \times 2 \times 3 = 6$
...
$n! = 1 \times 2 \times 3 \ldots n$
For convenience, zero factorial is defined as unity. Thus $0! = 1$.

Fibonacci Numbers The sequence of numbers 1, 1, 2, 3, 5, 8, 13, 21, 34, ..., each of which is the sum of the two previous numbers.

Greatest Integer Function The greatest integer not greater than **x**; denoted [x]. Thus $[3.2] = 3$ and $[-1.5] = -2$.

Lucas Numbers The sequence of numbers 1, 3, 4, 7, 11, 18, ..., each of which is the sum of the two previous numbers.

Odd Number An integer that is not evenly divisible by 2; any number of the form $2n + 1$, where **n** is an integer. Thus, 1, 3, 5, 7, 9, ... are odd numbers.

Palindrome A word, verse, number of two or more digits, etc., that is the same when read backward or forward. The number 344575443 is a palindromic number.

Prime Number A counting number, larger than 1, that has no divisors other than itself and 1. The sequence of primes begins with 2, 3, 5, 7, 11, ...

Square Number Numbers that are the squares of integers such as 1, 4, 9, 16, 25, 36, 49, ...

Square Root The square root of a number **n**, denoted $\sqrt{n}$ or $\sqrt[2]{n}$ is the non negative number **a** such that $a \times a = n$. Examples: $\sqrt{16} = 4$; $\sqrt{-4}$ is not defined, because there is no positive number **a** such that $a \times a = -4$. Examples: $\sqrt{(-2)^2} = \sqrt{4} = 2$, $-\sqrt{9} = -3$, $\sqrt{(-3)^2} = \sqrt{9} = 3$, $\sqrt{0} = 0$.